The Art of Regression Modeling in Road Safety

Ezra Hauer

The Art of Regression Modeling in Road Safety

Springer

Ezra Hauer
Professor Emeritus
University of Toronto
Toronto, ON, Canada

Additional material to this book can be downloaded from http://extras.springer.com

ISBN 978-3-319-12528-2 ISBN 978-3-319-12529-9 (eBook)
DOI 10.1007/978-3-319-12529-9
Springer Cham Heidelberg New York Dordrecht London

Library of Congress Control Number: 2014953294

Printed on acid-free paper

Springer is part of Springer Science+Business Media (www.springer.com)

To Gordon Newell, mentor, in gratitude.

Preface

This book has two objectives. The first is to teach how to fit a multivariable statistical model to cross-sectional safety data using a simple spreadsheet. The second is to promote the understanding that is at the core of good modeling.

These twin objectives determine the flow and the structure of the book. After some preliminaries, a real data set is introduced. From here on and all the way to the last chapter, the same data are used to gradually build up a regression model. Along the way there are the "what and how" sections: what is an exploratory data analysis, how to use pivot tables, what is a curve-fitting spreadsheet and how to build one, how to use the Solver for parameter estimation, how to examine the quality of a fit by a CURE plot, what functions look like, etc. These are in support of the first objective, that of teaching how to fit a model to data. Interspersed between these are sections of a reflective nature. These support the "understanding" objective and speak about the "why, when, and whether" issues of modeling: why do we need to curve-fit, whether models be used in a cause–effect manner, when should a variable be added to the model equation, why it is important to know what function hides behind the data, etc.

Data about accidents, about the road, and about the traffic on it are routinely collected and maintained in databases. We know where reported accidents occurred and much about their circumstances. We also have information about many features of the road (grade, curve radius, lane width, speed limit, parking control, etc.) and of its traffic (daily volumes, percent of trucks, etc.). Inasmuch as these data pertain to what is observed to exist on a cross section of "units" (road segments, intersections, etc.) they are called "observational cross-sectional" data.

To be of use for evidence-based road safety management, the data need to be summarized and cast into the form of statistical models. Models serve three purposes:

1. To estimate how (un)safe are certain road segments, intersections, ramps, crossings, etc. thereby determining the size of the road safety problem that could perhaps be altered by interventions and design changes.

2. To estimate by how much safety has changed following an intervention or a design change.
3. To estimate by how much safety might be altered due to a change in some variable value.

The book focuses on the development of regression models for purposes 1 and 2. While the use of regression models for purpose 3 is commonplace, the trustworthiness of the result is in doubt. Still, it is possible that even this vexed purpose will be well served by the approach to modeling that is advocated here.

Who will use this book? Perhaps they will be graduate students with interest in road safety, perhaps professionals with responsibilities in data analysis, or perhaps others. I do not know how much math, probability, and statistics I can rely on. Some such knowledge is required by the very nature of the subject matter. There will be parts which some readers may find taxing. The hope is that judicious skipping will make the book accessible and useful for a variety of audiences. I tried to make the narration succinct; diversions, elaborations, and detail were relegated to footnotes. A glossary is provided to serve as a refresher of notation and acronyms; the index at the end is for finding the page on which a topic or concept is first introduced.

The book evolved from lecture notes used in a series of hands-on workshops and is richly laced with data-based illustrations, tables, and figures. The supporting materials are available for downloading. To access and download them, go to http://extras.springer.com/ and enter the ISBN of this book. The ISBN (International Standard Book Number) is found just after the title page. The information is in five folders: Data, PowerPoint presentations, Problems, Solutions, and Spreadsheets. Together with the book these materials will be of interest to the reader, student, and instructor.

The book, of course, can be read as is. However, unlike textbooks in the past, it can also be used actively and creatively. The reader is invited to use the downloadable materials to see how results were obtained; to modify, expand, and enrich the analysis; and to use the same spreadsheets with other data.

Modeling in this book is built around the use of Excel spreadsheets and readers are assumed to have facility in their use. Information about less commonly known spreadsheet functionalities is provided where needed. Of course there are specialized and sophisticated statistical software packages which, once acquired and mastered, will do a good, perhaps a superior, job of model development. However, I find that the spreadsheet provides all the essentials; it makes for intimate contact with the data, it has adequate and flexible visualization, the "pivot tables" serve for exploration, an optimization tool does the curve-fitting, and it is a hospitable environment for writing custom pieces of code.

Model development is often presented as if it was a nearly algorithmic sequence of steps, an ordered progression of activities from "Start" to "End." In my opinion such an approach tends to produce inferior results. It is better to think of model development as detective work with clues embedded in data. Like in a game of snakes and ladders, there are advances and setbacks whereby the modeler gradually moves towards a satisfactory outcome. Such work is well served by the atmosphere of a spreadsheet. For all these reasons, the spreadsheet is my environment of choice for both instruction and creative modeling.

The modeling approach described in these pages may be thought old-fashioned, perhaps unsophisticated. Emphasis is on what is of essence. Papers describing novel modeling ideas and newfangled statistical techniques are being published daily and I make no attempt to capture the latest. In defense I only say that the quality of a meal depends more on the skill of the cook and the time spent on its preparation than on the modernity of the food processor. As has been said: "... second-rate minds grappling with first-rate problems can teach you more than first-rate minds lost in the shrubbery." (Lilla, 2013, p. xii).

Toronto, ON, Canada Ezra Hauer

Reference

Lilla M in Foreword for Berlin I (2013) Against the current, 2nd ed. Princeton University Press, Princeton

Contents

Glossary

Notational Conventions

$\sigma\{.\}$	Standard deviation of what the dot stands for
^	A caret stands for "estimate of" the letter below
$E\{.\}$	Expected value of what the dot stands for
ln	Natural logarithm
$P(K=k)$ or $P(k)$	Probability that the random variable K takes on the value k

Acronyms

AADT	Annual average daily traffic
AIC	Akaike information criterion
BIC	Bayesian information criterion
C-F	Curve-fitting
CMF	Crash modification factor or function
EDA	Exploratory data analysis
F&I	Fatal and injury
ML	Maximum likelihood
NB	Negative binomial
NM	Negative multinomial
N-W	Nadaraya-Watson
OVB	Omitted variable bias
pdf	Probability density function
PDF	Probability distribution function
PDO	Property damage only
RHR	Roadside hazard rating
SD	Squared differences (residuals)
SPF	Safety performance function

SSD	Sum of squared differences (residuals)
TAB	Total accumulated bias
VBA	Visual basic for applications
VIEDA	Variable introduction exploratory data analysis
vpd	Vehicles per day

Greek

α, β	Parameters
β_0	Scale parameter
β_i	Regression parameters $i = 1, 2, 3, \ldots$
ε_{f,X_i}	Elasticity of function "*f*" with respect to change in variable X_i
θ	Gamma distributed random variable with mean = 1 and variance = $1/b$
λ	Mean number of reported accidents per unit of time
μ_i	Expected number of accidents of unit i
$\sigma\{\mu\}$	Standard deviation of the μ's in a population of units
$\hat{\sigma}^2(i)$	Sum of sorted squared residuals from 1 to i
$\pm\hat{\sigma}'(i)$	Standard error for CURE plot at index = i

Latin

a, b	Parameters of the gamma and the negative binomial distributions
$A, B, \ldots$	Traits
c	Vehicle "concentration"—the (average) number of vehicles/(lane-km)
$E\{\mu\}$	Expected value of the μ's in a population of units
f	The expected number of reported accidents per lane per second
$f()$	The function linking predictor variables and parameters
h	Bandwidth in the N-W nonparametric regression
$\bar{h}$	Average headway
i	Counter (index) of units
j	Counter (index) of time periods
$K(.)$	Kernel function in the Nadaraya-Watson nonparametric regression
K, k	Count of accidents and a certain value of that count
$\mathcal{L}(.)$	Likelihood. The dot in the parenthesis is a placeholder for parameters.
L_i	Length of road segment i
$\ln(\mathcal{L}^*)$	Abridged log-likelihood
m	As subscript denotes "multivehicle"

N	Number of bins or levels of a trait
n	Number of units or, occasionally, of accident counts
NB	Negative binomial
NM	Negative multinomial
p	Probability of a vehicle to be in a crash in the next second
$P(.)$	Probability of the event that the dot stands for
r	The probability of the crash to be reported
s	Standard error or, occasionally, the sum of accident counts
$_{s}$	As subscript denotes "single-vehicle"
T	Duration of a time period
$V\{.\}$	Variance of the random variable that the dot stands for
$\overline{v}$	Average speed
v_0	Average free-flow speed
v_b	Average speed at which a bottleneck begins to form
$X_1, X_2, \ldots$	Variables in model equation
Δt	Small time interval

What Is What

1

Abstract

Statistical models that use data to express the safety of populations of units (road segments, intersections, grade crossings, etc.) as a function their traits (traffic, geometry, operation) are nowadays called Safety Performance Functions; SPFs for short. To make this notion precise one has to say what is meant by "safety," "unit," "population," and "trait." Most importantly, one has to be clear about what exactly is the information that a SPF provides and what are its practical uses. These are then illustrated by a series of examples. The practical uses of SPFs call for an approach to modeling which is somewhat different from what is usually done.

1.1 Units and Their Safety Property

What should one call the safety of Bobcaygeon Road between the Scotch Line and Plantation Roads in Ontario or the safety of the intersection of Eglinton and Don Mills roads in Toronto? These questions refer to safety as a property of some elements of the real world to be called "units." A road segment, an intersection, a vehicle, or a person is a "unit." A key feature of a "unit" is that it may be involved in accidents[1] (crashes) or that crashes (accidents) may occur on it. While the count of

[1] Those who prefer to use "crash" instead think that "accident" has connotations of being unavoidable, without cause, and thus unpreventable. This, they fear, might weaken the resolve to reduce crashes and their harm. Since the very purpose of studying road safety is to assist in the task of managing the frequency and severity of accidents, such an interpretation makes no sense in this book. In this book "crash" and "accident" describe the same event. One reason for not shunning the term "accident" is that it is the common currency in the community of transportation professionals. Another reason is that the word "accident" provides the proper associations for the randomness inherent in accident counts. The editor of the Canadian Oxford Dictionary (Barber 1999) says that: "No dictionary that I know of uses the word "unpreventable" in any of its definitions of the word "accident." Were they to do so, the definition would be inaccurate and

E. Hauer, *The Art of Regression Modeling in Road Safety*,
DOI 10.1007/978-3-319-12529-9_1

accidents is an indicator of the safety of a unit, it is not identical to it. The safety property of a unit is defined to be *the number of accidents by type and severity, expected to occur on or to it in a specified period of time.*[2]

Two elements of this definition need clarification. First, the word "expected" usually corresponds to "average in the long run." This works well for the idealized devices used in the instruction of probability theory: coin tosses, decks of unmarked cards, urns with balls, and fair dies. It works well because all these devices can be plausibly assumed not to change with repeated use. In contrast, the safety of real units changes with time.

Accordingly, one has to interpret term "expected" by a conditional and counterfactual statement as "what the limit of the long-term average would be if[3] it was possible to freeze all the safety-related traits and circumstances of the unit."

The second element in the definition of safety which requires clarification is the phrase "by type and severity." This means that, generally, the safety property of a unit is not a single number. Thus, for example, the safety of the intersection of High and Main streets in the 2-year period 2012 and 2013 could be described by the array in Table 1.1.

From such an array one can obtain the row sums (5.00 rear-end, 2.40 angle, 0.42 single-vehicle, and 0.08 pedestrian accidents), column sums (4.80 PDO, 2.75 injury, and 0.35 fatal accident), and the total (7.90 accidents). Each accident may involve one or more drivers and their vehicles. Therefore for multi-vehicle categories such as rear-end and angle accidents the additional categorization by number of involved vehicles may be needed.[4]

---

not reflect the actual usage of the word. The defining terms that dominate are "unexpected," "unforeseen," "unintentional," and "undesirable . . . ." Most people recognize that the things we refer to as "accidents" do indeed have causes, whether it be an unplanned pregnancy, slipping on a banana peel or the dog peeing on the rug and that accidents are preventable." In some cases, such as when one speaks of Crash Modification Functions (CMFs) the use of "crash" is so well established that its use already seems natural. However, the exclusive use of "crash" is often a sign of advocacy, an impression I want to avoid. This is why both terms will be used interchangeably.

[2] For a more detailed discussion of how safety is defined see Chap. 3 in Hauer (1997).

[3] As if the notion of "average in the long run" was not enough of an obstacle, the "would be if" phrase further distances the safety property from what is observable. Not only is safety not the same as the number of accidents, it is now something that can be imagined and perhaps estimated but can never be observed.

[4] Alternatively, one may want to speak not of the number of accidents but of the number of accident-involved vehicles or drivers. Similarly, an injury accident is one in which one or more persons were injured. Thus one may wish to describe the safety of a unit not by the number of injury accidents but by the number of injured persons. To convert from "number of accidents" to "number of involved vehicles or drivers" one needs to know the ratio "involved vehicles/accident." For the US from 1988 to 2005 the ratio remained steady at 1.724 ± 0.015. To convert from "fatal accidents" to "persons killed" knowledge of the ratio "persons killed/fatal accident" is needed. For the US from 1988 to 2005 the ratio for fatalities/fatal crash was 1.113 ± 0.004 which is similar to that in Michigan (1.104 ± 0.013 and 1.114 ± 0.019 for those who had been drinking), in Ohio (1.088 ± 0.013), Wisconsin (1.130 ± 0.022). It is also similar to the ratio pedestrian fatalities/ Pedestrian Fatal accidents which in Michigan was 1.080 ± 0.023.

Table 1.1 Expected accidents

	Accident severity		
Accident type	PDO	Injury	Fatal
Rear-end	3.10	1.70	0.20
Angle	1.40	0.90	0.10
Single-vehicle	0.30	0.10	0.02
Pedestrian		0.05	0.03

To save on words the symbol μ will be used to denote the expected number of accidents of a certain unit; subscripts will provide detail when necessary. Thus, for example, in Table 1.1, $\mu_{\text{high and main, rear-end, fatal, 2012 and 2013}} = 0.20$ accidents. A μ always pertains to a certain unit and time period.

1.2 Safety, Traits, and Populations

That the safety (μ) of a unit is determined by its many safety-related traits will be taken as a foundational axiom[5]. Units with identical safety-related traits have the same μ. A trait is said to be "safety related" if when it changes the μ of the unit changes.

The cube in Fig. 1.1 represents a unit, a road segment, the safety of which is μ. The μ of this road segment is determined by groups of factors. Invoking associations from heredity, these are shown as "safety chromosomes"; there is a chromosome for "climate," another for "geometry," etc. Carrying the analogy a step further, each chromosome consists of a multitude of traits depicted via the DNA imagery. Thus the "geometry" chromosome of a road segments has a group of traits pertaining to "lanes," a group of traits for "vertical alignment," etc. If the unit was an intersection one would have chromosomes or traits for "number of approaches," "type of traffic control," etc. If the unit was a person there would be a chromosomes and traits to capture personality, age, gender, etc.

At times it is the safety of a set of units which is of interest. Thus, for example, one might want to know the safety of rural two-lane roads in Ontario, or of signalized intersections in Toronto, or of the cohort of 75–85 years old Canadian drivers. In such cases we speak of the safety of a population of units. *Units that share a number of traits form a population*[6].

To illustrate consider the population the units of which are segments of rural two-lane roads in Colorado. Their shared traits are: (1) State: Colorado, (2) Road Type: two-lane, (3) Setting: rural. Each trait consists of a couplet – the name of the

[5] As always, an axiom is a starting point of reasoning, a premise that is accepted as true without proof and without controversy. Inasmuch as this book is about models in which the safety of units is to be represented as a function of their traits this is obviously a foundational axiom.

[6] In his foundational paper Fisher (1922, p. 311) defines the principal task of statistics to be "the reduction of data," and states that "this object is accomplished by constructing a hypothetical infinite population, of which the actual data are regarded as constituting a sample."

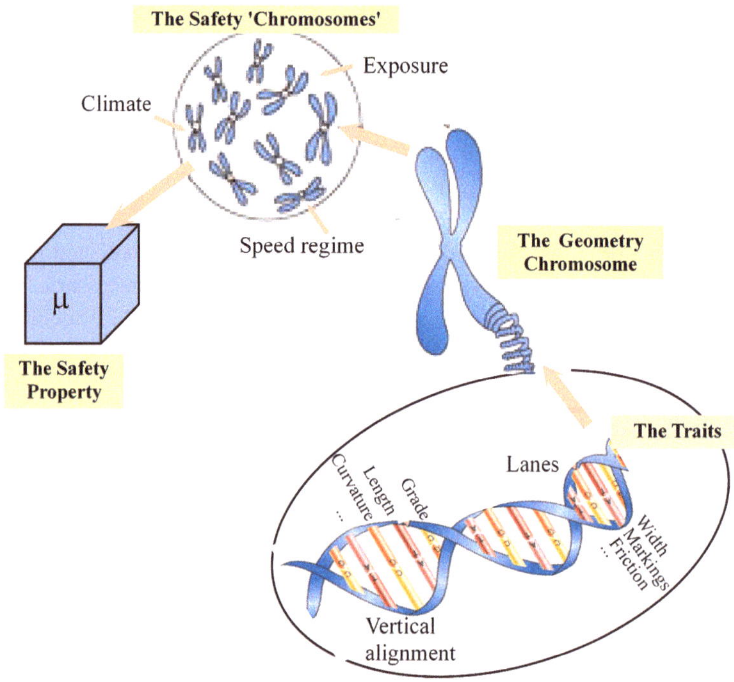

Fig. 1.1 The safety property (μ) of some road segment as function of traits

trait and its "level." Thus, when the trait name is "State" the trait level could be Alabama, Alaska, Arizona, etc.; when the trait name is "Road Type" the level could be four-lane undivided, six-lane access controlled, etc.

Adding a trait to those already shared by a population of units defines a new population, a subpopulation of the original one. Thus, for example, adding trait (4) Segment Length: 2.5–3.5 miles forms a subpopulation of the population defined by the aforementioned traits (1)–(3); adding to these trait (5) AADT[7]: 1,000–2,000 vehicles defines still another subpopulation, etc.

The more numerous are the traits which define a population the fewer are the units that belong to it. To illustrate, in Sect. 3.2, I will introduce data about 5,323 road segments sharing traits (1)–(3). The subpopulation defined by adding trait (4) contains 597 road segments, and when trait (5) is added the resulting subpopulation has only 119 road segments. There comes a point when the units are so few that it makes no sense to speak of a real population. Thus, for example, even with the rich Colorado data there will be no units in a population defined by a Segment

[7] Annual average daily traffic.

Length bin of 2.6–2.7 miles and an AADT bin of 2,400–2,420 vehicles. One can still meaningfully ask what would be the mean of the μ's of units with these traits. However, the population of units is now an imagined one.

While the units in a population, whether real or imagined, share a number of safety-related traits, they inevitably differ in many other safety-related traits. Thus, for example, the Colorado road segments which share traits (1)–(5) will still differ in, curvature, slope, cross section, speed distribution, proportion of trucks, proximity to hospital, etc. It follows from the foundational axiom that for any population of units, no matter how many traits are used to define it, the only realistic point of departure is to assume that the μ's of its units are all different. This is shown schematically in Fig. 1.2. While all the units share the trait of being "cubes," they differ in other traits (size, outline, shading, etc.) and therefore each cube, each unit, has its distinct μ.

To describe the safety of a population with parsimony I will use the *mean* and the *standard deviation* of the μ's of its units. These are the two quantities used in practical applications. The usual notation for the mean of the μ's is $E\{\mu\}$ and for the standard deviation of the μ's it is $\sigma\{\mu\}$. The square of $\sigma\{\mu\}$ is the variance of the μ's denoted as $V\{\mu\}$.

Notation to get used to:
μ – Expected number of accidents for a unit
$E\{\mu\}$ – Mean of the μ's of units in a population
$\sigma\{\mu\}$ – Standard deviation of the μ's of units in a population
$V\{\mu\} = \sigma^2\{\mu\}$
$\hat{\square}$ – Estimate of whatever is in the box

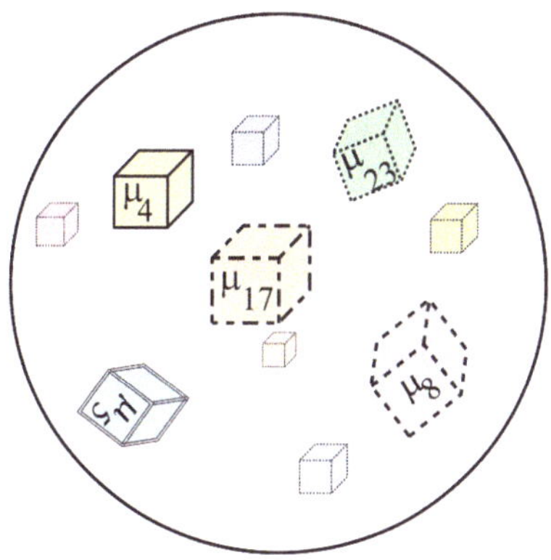

Fig. 1.2 A population of units that have some but not all safety-related traits in common

Notation of this kind may be off putting to non-statisticians. However, once understood and remembered its use narrows the scope for misunderstanding.[8] "E" will always stand for "expected value," "σ" for "standard deviation," V for variance, and the dot in $\{.\}$ for whatever variable the E, σ, or V apply to.

I have to test the patience of the non-statistician a bit further. The μ of units or the $E\{\mu\}$ and $\sigma\{\mu\}$ of populations can never be known; they can only be estimated with various degrees of accuracy. A caret ^ above a letter will stand for "estimate of." Thus, for example, $\hat{\mu}_i$ is the estimate of the μ of unit i; the standard deviation of $\hat{\mu}_i$ is, $\sigma\{\hat{\mu}_i\}$, and its estimate, the so called *standard error*, is $\hat{\sigma}\{\hat{\mu}_i\}$, etc. This completes the necessary notational conventions.

The preceding elaborate definitional and notational arsenal was needed lay down a clear and concise definition of the SPF:

A Safety Performance Function (SPF) is a device which for a multitude of populations provides estimates of two elements:

1. $E\{\mu\}$, *the mean of the* μ*'s in populations*;
2. $\sigma\{\mu\}$, *the standard deviation of the* μ*'s in these populations.*

The estimates of $E\{\mu\}$ and $\sigma\{\mu\}$ which an SPF provides are needed for the practice of road safety management. Their uses are discussed next.

1.3 What $\hat{E}\{\mu\}$ and $\hat{\sigma}\{\mu\}$ Are Needed for

It is important to say why we need the estimates of $E\{\mu\}$ and $\sigma\{\mu\}$[9] for the practice of road safety management. The reasons are several and fall into the two groups shown in Table 1.2.

In the first group are occasions when the $E\{\mu\}$ itself is of interest. Thus, for example, we may want to know what is the "normal" safety of units with some traits in order to judge whether a certain unit with the same traits is "deviant" or "unsafe" and may require attention, perhaps treatment. The related activity is called "blackspot identification" or "network screening[10]." Another occasion in this group of reasons is when we want to know how different is the $E\{\mu\}$ of one population of units from that of another population. Thus, for example, one may ask how different are $E\{\mu\}$'s of two populations of road segments that have the same known traits but differ in whether their shoulders are paved or unpaved. Such a difference might indicate what might be the average safety effect of shoulder paving.

[8] As Whitehead (1958) pointed out: "By relieving the brain of all unnecessary work, a good notation sets it free to concentrate on more advanced problems, and in effect increases the power of the race."

[9] Much of the highway safety manual (AASHTO 2010) is devoted to SPFs and their practical uses.

[10] AASHTO. Highway safety manual (2010), Chap. 4.

Table 1.2 The practical tasks for which estimates of $E\{\mu\}$ and $\sigma\{\mu\}$ are needed

Group I: Focus is on the $E\{\mu\}$ of populations of units	Group II: Focus is on the μ of specific units
What is normal for units with given traits?	Is this unit a "blackspot" or "unsafe"?
How different are the $E\{\mu\}$'s of two populations?	What might be the safety benefit of treating this unit or changing its design?
	What was the safety effect of some treatment?
To answer we need estimates of $E\{\mu\}$.	To answer we need estimates of $E\{\mu\}$ and of $\sigma\{\mu\}$.

To the second group of reasons belong those occasions when we have to estimate the μ of a specific unit and $E\{\mu\}$ helps with this task. This is the case when the unit has no accident history or when the μ of a unit is estimated by the Empirical Bayes method which mixes the estimate of $E\{\mu\}$ with the accident history of the unit [11]. We need to estimate the μ of specific units when we ask whether a certain road segment is a "blackspot," when the aim is to estimate what might be the safety benefit of a treatment, and when inquiring by how much some treatment has changed the μ of treated units. On all these occasions in which the μ of specific units is of interest, one has to have an estimate of both $E\{\mu\}$ and $\sigma\{\mu\}$.

It should be clear that to serve these uses many SPFs are needed. Thus, for example, if paving shoulders on two-lane rural roads is thought to affect mainly run-off-the-road accidents, then an SPF for this accident type and road is needed. If blackspots involving left-turns at signalized intersections are of interest, the corresponding SPF is needed. Thus, not only are the practical tasks many and diverse, so are also the SPFs defined by unit and accident type. It follows that the task of SPF development is and will continue to be a necessary accompaniment of quantitative road safety management. The numerical examples to follow illustrate some of the tasks of quantitative road safety management.

1.4 How $\hat{E}\{\mu\}$ and $\hat{\sigma}\{\mu\}$ Are Used: Numerical Examples

So as not to stray too far from our subject matter but to still link the definition of the SPF to how its products might be used, this section provides a sequence of illustrative examples. To keep the section brief, the examples are kept simple.

[11] The Highway Safety Manual, the Interactive Highway Safety Design Model software and the SafetyAnalyst software all make use of the Empirical Bayes method. For a brief description see Section 3.5 of AASHTO (2010).

1.4.1 Data for Two Populations

The linked illustrations to follow make use of data about two populations of units. One pertains to drivers, the other to road segments. The data in Table 1.3 is about persons licensed to drive in Connecticut in the 6-year period 1931–1936.[12] The data in Table 1.4 is about rural two-lane roads in Colorado.[13]

In the computations to follow details will be provided for the driver data while, to avoid repetition, for the road segment data only the end-results will be given.

1.4.2 Estimating $E\{\mu\}$ and $\sigma\{\mu\}$

What is needed in the various applications to follow are the estimates of the mean of μ's and of their standard deviation, the $\hat{E}\{\mu\}$ and $\hat{\sigma}\{\mu\}$. Here the question is how these estimates can be extracted from the data in Tables 1.3 and 1.4. In these tables k denotes a accident count and $n(k)$ the number of units in the population, either drivers or road segments, with k accidents. Computing the sample mean and sample variance of the accident counts is simple.

Table 1.3 Connecticut drivers (1931–1936)[14]

Number of accidents, k	Number of drivers, $n(k)$
0	23,881
1	4,503
2	936
3	160
4	33
5	14
6	3
7	1
Total	29,531

[12] The data are from Cobb (1940). There is a certain attraction in the use of venerable old data.

[13] The data are about 5,323 road segments of varying length totaling 6,029 miles, with an AADT up to about 20,000 (average 2,151), on which, during 13 years, 21,718 injury and fatal and 52,317 Total accidents were recorded. The data will be introduced in Sect. 3.2 and then used throughout the book.

[14] To access and download data go to http://extras.springer.com/ and enter the ISBN of this book. Look in the "Data" folder for "2a. Connecticut drivers.xls," or "2b. Connecticut drivers.xlsx"

Table 1.4 Fatal and injury accidents on Colorado road segments that all have a 13 years accident history and are 0–1 miles long[15]

Number of accidents, in 5 years, k	Number of segments, $n(k)$
0	1,499
1	777
2	398
3	273
4	141
...	...
30	2
37	1
48	1
Total	3,516

$$\text{Sample mean} = \hat{E}\{k\} = \sum_{k=0}^{\text{largest } k} k \frac{\text{Number of units with } k}{\text{Number of units}} \tag{1.1}$$

$$\text{Sample variance} = \hat{V}\{k\} = \sum_{k=0}^{\text{largest } k} \left(k - \hat{E}\{\mu\}\right)^2 \frac{\text{Number of units with } k}{\text{Number of units}} \tag{1.2}$$

As shown in Fig. 1.3, k is not μ; k is an observable accident count (an integer), while μ is its expected value (a real number), a quantity that can be estimated but not directly observed. The μ's are represented by circles and the k's by squares. Thus, for example, while the μ of unit 1 is shown as 0.71 accidents, the count of accidents on unit 1 in the same time period was 3. The arrow linking the circles and the squares represent the randomness in data generation process.[16]

While it is true that the sample mean of accident counts $\hat{E}\{k\}$ is an unbiased estimate of $E\{\mu\}$ and therefore one may use $\hat{E}\{\mu\} = \hat{E}\{k\}$, it is also true that the variance of accident counts $V\{k\}$ is always larger than the $V\{\mu\}$ and cannot serve as its unbiased estimator.[17] One way[18] to estimate the $\sqrt{V\{\mu\}} = \sigma\{\mu\}$ is by:

---

[15] To access and download data go to http://extras.springer.com/ and enter the ISBN of this book. Look in the "Data" folder for "3a. Colorado road segments, k and $n(k)$" or "3b. Colorado road segments, k and $n(k)$.xlsx"

[16] It is commonly assumed that the k's are Poisson distributed with μ as mean. This corresponds to the specific data generation process in which events are rare and the occurrence of one event is not influenced by the occurrence of another one. This process is described more fully in Appendix A.

[17] The assertion that $V\{k\} > V\{\mu\}$ can be supported by visual intuition in Fig. 1.3; the squares will tend to be more widely dispersed than the circles. This must be so since the circles on the right will occasionally generate some squares even further out to the right and the circles on the left will occasionally generate squares even further left.

[18] The derivation of this formula is in the Appendix C. Another way to estimate $\sigma\{\mu\}$ will be described later in the context of maximizing likelihood. Since the variance cannot be negative, it makes sense to replace negative values of $\hat{V}\{k\} - \hat{E}\{\mu\}$ by 0. However, the estimator in (1.3) is biased. To see why, consider the limiting case when all μ's in a population are the same. Repeated estimation by (1.3) would yield estimates of $\sigma\{\mu\}$ that are either positive or 0.

$$\hat{\sigma}\{\mu\} = \begin{cases} \sqrt{\hat{V}\{k\} - \hat{E}\{\mu\}} & if\ \hat{V}\{k\} > \hat{E}\{\mu\} \\ 0 & \text{otherwise} \end{cases} \tag{1.3}$$

How $\hat{\sigma}\{\mu\}$ can be computed from the data in Table 1.3 is shown in Fig. 1.4.[19]

The summands for (1.1) are in column D and their sum is in cell D11. Thus, for the population of Connecticut drivers the estimate of the mean of μ's is 0.240 accidents in 6 years. The summands for (1.2) are in column E and their sum in cell E11. Using (1.3), $\hat{\sigma}\{\mu\} = \sqrt{0.306 - 0.240} = \pm 0.256$ accidents in 6 years. For the Colorado road segment data in Table 1.4 the corresponding estimates are $\hat{E}\{\mu\} = 1.86$ and $\hat{\sigma}\{\mu\} = \pm 2.84$ F&I accidents in 13 years. Their computation is left as an exercise.

The estimates of $\sigma\{\mu\}$ are a quantitative expression of what is obvious. Namely, that not all Connecticut drivers and not all Colorado road segments had the same μ. The reason is that drivers differ in "gender," "age," "annual mileage," and many other safety-related traits, and that road segments differ in amount of traffic, terrain, curvature, etc. However, the purpose was not to demonstrate the obvious. The estimates of $E\{\mu\}$ and $\sigma\{\mu\}$ are needed for the practical applications illustrated below.

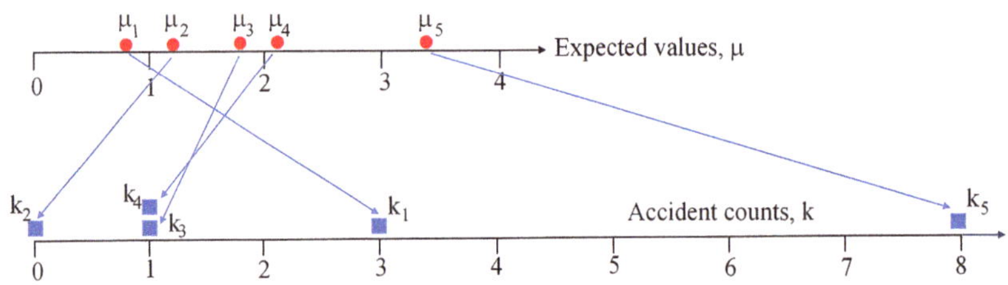

Fig. 1.3 Expected values and accident counts

Fig. 1.4 Sample mean and sample variance of accident counts and the estimate of $\sigma\{\mu\}$

	A	B	C	D	E
1	**Data**			**Computations**	
2	**k**	**n(k)**	**B/B$11**	**A * C**	**(A-D$11)²*C**
3	0	23881	0.8087	0.000	0.047
4	1	4503	0.1525	0.152	0.088
5	2	936	0.0317	0.063	0.098
6	3	160	0.0054	0.016	0.041
7	4	33	0.0011	0.004	0.016
8	5	14	0.0005	0.002	0.011
9	6	3	0.0001	0.001	0.003
10	7	1	0.0000	0.000	0.002
11	**Sums**	**29531**	**1.0000**	**0.240**	**0.306**
				$\hat{E}\{k\}$	$\hat{V}\{k\}$

[19] To access and download the spreadsheet with these and subsequent computations go to http://extras.springer.com/ and enter the ISBN of this book. Look in the 'Spreadsheets' folder for 'Chapter 1. Connecticut drivers.xls or .xlsx'.

1.4.3 How Many High-μ Units Are There?

One common application is "screening,"[20] be it of drivers or road segments. The aim is to identify units with an unusually high μ in the hope that the safety of such units can be efficiently improved. Depending on context, these units are called blackspots, sites with promise, accident prone drivers, unsafe vehicles, etc.

To design a sensible screening tool one should begin by forming an idea about how prevalent are the high-μ units in a population. Thus, for example, one could ask what proportion of drivers or road segments had a μ that is, say, five times the population average? For the Connecticut drivers that average was 0.240 accidents in 6 years and so the question here is how many of the 29,531 Connecticut drivers are expected to have a μ larger than, say, $5 \times 0.24 = 1.2$ accidents in 6 years? For the Colorado road segments the average was 1.85 and the question is how many of the 3,516 segments are expected to have a $\mu > 5 \times 1.86 = 9.3$ accidents in 13 years?

To answer these questions one must make an assumption about the form of the probability distribution function of the μ's in the population. A popular assumption is that the μ's are Gamma distributed.[21] The Gamma distribution has two parameters, a and b which are related to $E\{\mu\}$ and $V\{\mu\}$ as shown in (1.4).

$$E\{\mu\} = \frac{b}{a},\ V\{\mu\} = \frac{b}{a^2} \text{ and therefore } a = \frac{E\{\mu\}}{V\{\mu\}},\ b = aE\{\mu\} \tag{1.4}$$

With this assumption one can compute the probability that amongst the Connecticut drivers $\mu > 1.2$ as shown in Fig. 1.5.

The estimates of $E\{\mu\}$ and $\sigma\{\mu\}$ obtained earlier in Sect. 1.4.2 are in cells B3 and B4. The computed values of a and b in cells E3 and E4 are based on (1.4). The Excel function GAMMADIST(μ, b, $1/a$, true) in cell B7 and below gives the cumulative probability of μ being less than the value in column A of the same row. Thus, as shown in cell B31, if the "Gamma assumption was a good approximation, 0.9906 % of these drivers were expected to have $\mu \le 1.2$ accidents in 6 years. Since there are 29,531 drivers in this population, about $29{,}531 \times (1 - 0.9906) = 278$ can be expected to have had $\mu > 1.2$ accidents in 6 years.

For the population of road segments in Table 1.4 the corresponding estimates of a and b are 0.230 and 0.427; for these the Gamma cumulative probability distribution function is in Fig. 1.6. The probability that the μ of a road segment in this population exceeded $5 \times 1.86 = 9.30$ F&I accidents in 13 years is about 3.1 %. Thus, of the 3,516 road segments about 108 are expected to have a μ which is more than five times the population mean.

[20] The Free Dictionary defines screening as: "The initial evaluation of an individual, intended to determine suitability for a particular treatment . . . ." When screening is applied to transportation infrastructure its role is "to identify and rank sites from most likely to least likely to realize a reduction in crash frequency. . ." (AASHTO 2010, p. 4.1).

[21] This is a flexible PDF, is available as a function in spreadsheets, and makes the algebra simple. As will be shown in Sect. 1.4.6, the Gamma assumption is often well supported by the data. More about this is in Appendix D.

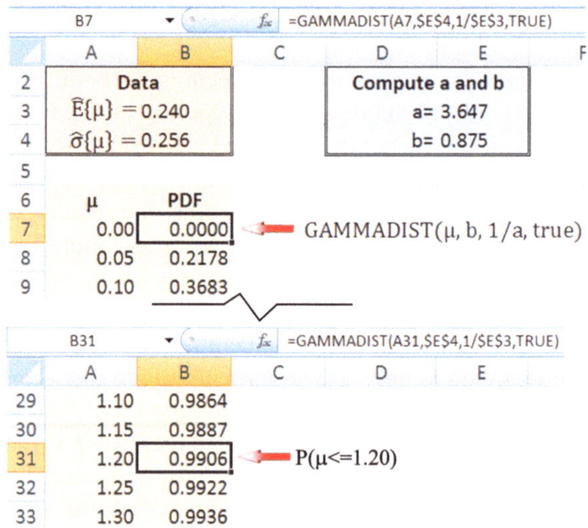

Fig. 1.5 Computing values of the cumulative Gamma distribution for Connecticut drivers

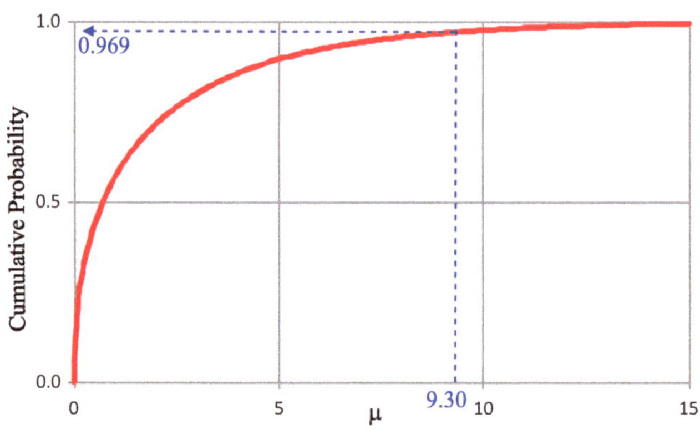

Fig. 1.6 The PDF of μ for the population of road segments in Table 1.4

It is the knowledge of both $E\{\mu\}$ and $\sigma\{\mu\}$ that allowed us to form an idea about how prevalent or scarce are the high-μ units in a population, be these road segments or drivers. How well a screen that is based on the accident history of units is likely to perform will be examined next.[22]

[22] Even without any computation one can say that when $\sigma\{\mu\}$ is small compared to $E\{\mu\}$ the proportion of high-μ units in the population is bound to be small. In that case their identification is difficult and perhaps unprofitable. Conversely, when $\sigma\{\mu\}$ is large compared to $E\{\mu\}$ the proportion of high-μ units is sizeable and therefore they will be easier to catch in the screening net. A numerical example may explain what "small" and "large" mean. Suppose that $E\{\mu\} = 1$ and that the aim is to identify units with $\mu > 4$. If $\sigma\{\mu\} = 1$ then, by the Normal "rule of thumb," one may expect 1 of 1,000 units to have a $\mu > 4$. However, if $\sigma\{\mu\} = 3$ then one may expect 160 of 1,000 units to have a $\mu > 4$.

1.4.4 The Performance of a Screen

Having established how many high-μ units there might be in a population, the next question is how well can be these high-μ units identified on the basis of their accident history. Specifically, how many of those Connecticut drivers or Colorado road segments that had many accidents also had a μ that is larger than five times their population average?

This answer too makes use of the estimates of $E\{\mu\}$ and $\sigma\{\mu\}$ or, more directly, of the "a" and "b" which are a function of $E\{\mu\}$ and $\sigma\{\mu\}$. It can be shown[23] that if the μ's in a population are Gamma distributed with parameters a and b then the μ's in the subpopulations of units that all recorded $k = 0, 1, 2, \ldots$ accidents are also Gamma distributed with

$$E\{\mu|k\} = \frac{k+b}{a+1} \quad \text{and} \quad V\{\mu|k\} = \frac{k+b}{(a+1)^2} \tag{1.5}$$

To examine these probability distributions the spreadsheet in Fig. 1.5 can be reused except that the "b" in the GAMMADIST function has to be replaced by $\mathrm{b}+k$ and the "a" by $a+1$. Thus, for example, as shown in the "formula bar" of Fig. 1.7, to obtain the PDF of μ's for drivers who recorded four accidents in 6 years, four is added to parameter the "b" from E4 and "1" is added to the "a" from cell E3.

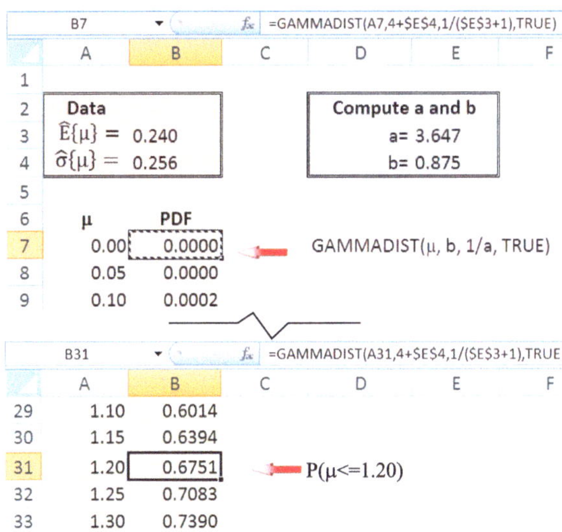

Fig. 1.7 The Gamma PDF for drivers with $k = 4$ accidents in 6 years

[23] See Appendix F and Hauer (1997, p. 192).

As shown in cell B31, some 68 % of the drivers with four accidents are expected to have $\mu \leq 1.20$ accidents in 6 years. Since there were 33 such drivers (Table 1.3), had these been selected for some treatment one would expect $33 \times (1 - 0.675) = 11$ to be "correct positives and the remaining 22 to be "false positives[24]." Proceeding similarly for $k = 5$, 6, and 7 makes for Table 1.5.

The problem of this crash-based screen is now obvious. If some remedial action was administered to drivers who recorded, say, four or more accident in 6 years then, of the 278 Connecticut drivers estimated to have $\mu > 1.20$ accidents in 6 years, only 21 would be caught in this net; 257 high-μ driver would be missed. Of the 51 drivers caught in the $k \geq 4$ net, 30 are expected to be "false positives."[25]

The corresponding information for the Colorado road segments with a screening net $k \geq 8$ accidents in 13 years is as follows: Of the 3,516 segments in Table 1.4 about 108 are expected to have $\mu > 9.3$ accidents in 13 years. Caught in the net are 172 segments of which 81 are expected to be correct positives (and 92 false positives). The difference of $108 - 81 = 27$ segments is the expected number of false negatives.

The performance of every kind of screening program, whether one seeks to identify drivers or elements of infrastructure, needs to be judged by these criteria: the number of correct positives, false negatives, and false positives. The quantification all three requires estimates of $E\{\mu\}$ and $\sigma\{\mu\}$.

So far the illustrations dealt with questions about a population of units; how many high-μ drivers are in it, how many false positives are in a subset, etc. To provide answers one had to have estimates of $E\{\mu\}$ and $\sigma\{\mu\}$. The next rung on the illustration ladder is about the safety of specific units.

Table 1.5 Connecticut drivers; correct and false positives

K	$n(k)$	$P(\mu \leq 1.2)$	False positives	Correct positives
4	33	0.68	22	11
5	14	0.51	7	7
6	3	0.34	1	2
7	1	0.21	0	1
Sums	51		30	21

[24] In the present illustration a unit is a "correct positive" if on the basis of its crash history it is identified as being "high-μ" when its μ is truly higher than five times the population mean. If the μ of the unit with $k = 4$ is smaller than five times the population mean, it is called a "false positive." In this terminology the word "positive" was inherited from medical phrase "testing positive." Correct positives are sick persons correctly diagnosed as sick; False positives are healthy people incorrectly identified as sick. A person testing positive for X who is determined to indeed have X, is a correct positive. In medicine it is nowadays common to speak about "Sensitivity" and "Specificity." Sensitivity is the (Number of persons identified as sick)/(Number of sick persons in population). In the Connecticut drivers example it is $21/278 = 0.08$. Specificity is the (Number of persons correctly identified as not sick)/(Number of well persons in the population). In the same example specificity is $(29{,}531 - 278)/(29{,}531 - 21) = 0.99$.

[25] More about this is in Hauer et al. (1993).

1.4.5 Estimating the μ of a Unit

The question about the safety of specific units is center-stage in two settings. First, one often asks how many accidents will be saved by treating a unit or set of units. In this setting the treatment is associated with a known "Crash Modification Factor (CMF)."[26] The safety benefit of implementing a treatment is μ(1-CMF). Thus, to know what benefit to expect by treating a unit or set of units one must have an estimate of their μ. The second common setting is when one wishes to assess what indeed was the safety benefit of treating some units. To do so one has to estimate the μ of the treated units both before and after the treatment.

To illustrate the first setting assume that a treatment with CMF = 0.95 is considered for those 51 Connecticut drivers with $k \geq 4$ accidents in 6 years. Consider first those with $k = 4$ accidents. By (1.5), the estimate of $E\{\mu|k = 4\}$ is (0.875 + 4)/(3.647 + 1) = 1.05 accidents in 6 years. The results of the same kind of computation for $k = 5$, 6, and 7 are shown in column F of Fig. 1.8.

Column G gives estimates of the number of accident expected by the $n(k)$ treated drivers. Summing over k from 4 to 7, one should expect these drivers 51 to have 58.4 accidents in 6 years. If so, the expected benefit of treating them is a reduction of $58.4 \times (1 - 0.95) = 2.92$ accidents in 6 years.

From here there is a seamless transition to the second setting, that of estimating the safety effect of a treatment. Now the story line is that a treatment has been applied to the Connecticut drivers with $k \geq 4$ in the 1931–1936 periods and the key question is how many accidents should one expect these drivers to record if the treatment had no effect and if there was no change in any safety-related conditions. The answer is the same as before: one should expect them to have 58.4 accidents in 6 years.

F7 =($E$4+D7)/($E$3+1)

	D	E	F	G	H	I
1						
2	Compute a and b		$a = \frac{E\{\mu\}}{V\{\mu\}}, b = aE\{\mu\}$			
3	a= 3.647					
4	b= 0.875		$E\{\mu\|k\} = \frac{k+b}{a+1}$ and $V\{\mu\|k\} = \frac{k+b}{(a+1)^2}$			
5						
6	k	n(k)	(b+k)/(a+1)	E*F		
7	4	33	1.05	34.6		
8	5	14	1.26	17.7		
9	6	3	1.48	4.4		
10	7	1	1.69	1.7		
11				58.4		

Fig. 1.8 Computing the number of accidents expected

[26] Crash Modification Factor or Function ≡ CMF ≡ (Expected accidents with the treatment implemented)/(Expected accidents if treatment is not implemented). CMFs for various treatments and interventions are listed in manuals, handbooks and available from public databases.

The reader may ask how come that the 51 drivers who together recorded 227 accidents in 6 years are expected to have only 58.4 accidents? The difference between 227 and 58.4 is, in this case, due to the regression-to-the-mean phenomenon. Indeed, the first expression in (1.5) is the Empirical Bayes[27] estimator designed to eliminate the regression-to-mean bias.

These simple illustrations serve to show that when one designs a screen to identify high-μ units, and when one tries to anticipate the benefit of a treatment, and also when one undertakes to evaluate the effect of a treatment, use is made of the estimates of $E\{\mu\}$ and $\sigma\{\mu\}$. These are the values which the SPFs provide.

1.4.6 Is the Gamma Assumption Sensible?

All these illustration as well as many of the procedures in the Highway Safety Manual (AASHTO 2010) hinge on the assumption that the distribution of μ's in a population of units can be adequately approximated by the Gamma distribution. Whether this assumption is justified can be examined only indirectly. That is, if the assumption was true, and if the count of accidents was well approximated by the Poisson distribution[28] then, the proportion of units with k accidents should obey the Negative Binomial (NB) distribution.[29]

$$P(k) = \frac{\Gamma(k+b)}{\Gamma(b)k!}\frac{a^b}{(a+1)^{k+b}} \tag{1.6}$$

Multiplying $P(k)$ by the number of units in the population yields the estimate of the number of units expected to have k accidents. These estimates can be compared to the observed number of units. Should the correspondence be good one may conclude that, for this specific population of units, the data do not negate the Poisson and Gamma assumptions.

For the two populations used in these illustrations, the correspondence between what was observed and what was estimated using the Poisson-Gamma assumptions is in Tables 1.6 and 1.7. The fit for the driver data is near perfect. The fit for the road segment data is much less impressive. Were one to conduct a formal χ^2 test of the

[27] The EB estimator is the linear combination $\alpha\, E\{\mu\} + (1-\alpha)k$ where the weight $\alpha = \frac{1}{1+V\{\mu\}/E\{\mu\}}$ (see Hauer 1997, p. 194). That this is the same as the expression in (1.5) can be shown by writing $E\{\mu|k\} = \frac{k+b}{a+1}$ as the sum $\frac{b}{a+1} + \frac{k}{a+1}$. After replacing b by $a\,E\{\mu\}$ this takes the form $\alpha\, E\{\mu\} + (1-\alpha)k$ where $\alpha = \frac{1}{1+1/a} = \frac{1}{1+V\{\mu\}/E\{\mu\}}$.

[28] See Appendix A.

[29] For more detail about the Negative Binomial distribution see Appendix D. The full notation in (1.6) would be $P(K=k|a, b)$ where K stands for the "Count of Accidents" and k for a specific value thereof. Usually the abridged notation $P(k)$ will be used.

Table 1.6 The fit of the NB for the Connecticut driver data

k	Observed $n(k)$	Fitted using negative binomial
0	23,881	23,891
1	4,503	4,497
2	936	907
3	160	187
4	33	39
5	14	8
6	3	2
7	1	0
Sum	29,531	29,531

Table 1.7 The fit of the NB for the Colorado road segment data

k	Observed $n(k)$	Fitted using NB	χ^2
0	1,499	1,718.0	27.9
1	777	596.5	54.6
2	398	346.0	7.8
3	273	227.6	9.0
4	141	158.6	1.9
5	117	114.2	0.1
...	...	...	...
13	10	12.8	0.6
14–15	16	17.8	0.2
16–17	6	10.9	2.2
≥18	18	14.7	0.8
Sum	3,516	3,512.5	112.1

hypothesis that segment counts come from this NB (Negative Binomial) distribution it would be clearly rejected.[30]

The question in the title of this section can now be answered. There is no reason why the μ's in populations of units should form a Gamma distribution. For some data the fit is almost magically good, for others it is poorer. It is good practice to check whether the data negates the Gamma assumption before conducting analyses that rely on it.

The sketch illustrations which make up Sect. 1.4 show how the estimates of $E\{\mu\}$ and $\sigma\{\mu\}$ which the SPF provides are used in various applied settings. Implicit in these illustrations is a point of view. Namely, that the purpose of SPFs is to provide estimates of $E\{\mu\}$ and $\sigma\{\mu\}$. As will become evident this point of view has implications for modeling. It is therefore, best to explain what the alternative point of view is.

[30] When the Gamma assumption is not supported by the data the numerical results based on it are in danger of being inaccurate.

1.5 The Chosen Perspective

The SPF is usually an equation on the left hand side of which are the estimates of $E\{\mu\}$ or $\sigma\{\mu\}$ and on the right hand side of which is some function of "traits" (variables) and of "parameters." This equation can be viewed from two different perspectives. From the perspective of the applications listed in Table 1.2 and the illustrations in Sect. 1.4, the model equation is thought to be a tool for generating estimates of $E\{\mu\}$ and $\sigma\{\mu\}$. This focus is shown in the upper part of Fig. 1.9.

The lower part of the figure shows the "research" perspective. When this point of view is selected one is interested in understanding how changing the various traits will affect $E\{\mu\}$ in order to predict the safety effect of design choices and interventions. From this perspective the focus is on the value of the unknown parameters and on the function "*f*" which links the traits and parameters. When viewed from the research perspective the SPF represents the current understanding of cause and effect and is thought to be a source of Crash Modification Factors and Functions.[31]

How one goes about developing an SPF depends to some extent on the chosen perspective.[32] In this book it will be assumed that the SPF is developed to support the practical application listed in Table 1.2. Therefore the aim will be to obtain good

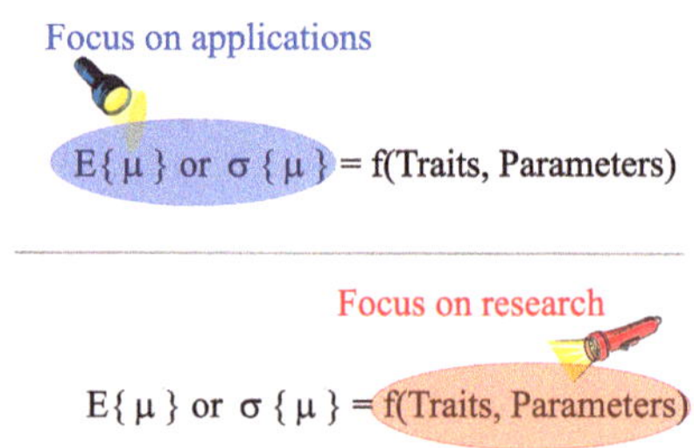

Fig. 1.9 Two perspectives on SPFs

[31] Opinions differ on whether such cause-and-effect interpretations of regression models are trustworthy. I am a cause-effect skeptic and believe that the use of SPFs as a source of crash modification factors or functions is fraught with problems. More on this is in Sect. 6.7.

[32] This point was made in Lehmann's important paper (Lehmann 1990, p. 163) where following the pioneering work of Neyman and Box the two perspectives on modeling are called "empirical" and "explanatory." When discussing the purpose of models Lehman says that "... the two kinds of models differ in their basic purpose. Empirical models are used as a guide to action, often based on forecasts of what to expect from future observations.... In contrast, explanatory models embody the search for the basic mechanism underlying the process being studied; they constitute an effort to achieve understanding." As noted later in Sect. 6.7.4, explanation and understanding go hand-in hand with cause and effect. Furthermore, "An explanatory model, as is clear from the very nature of such models, requires detailed knowledge and understanding of the substantive situation that the model is to represent. On the other hand, an empirical model may be obtained from a family of models selected largely for convenience, on the basis solely of the data without much input from the underlying situation" (p. 164).

estimates of $E\{\mu\}$ and $\sigma\{\mu\}$. This aim does not always coincide with the focus on parameter estimates which characterizes the "research" perspective.

1.6 Summary

Units are some elements of the real world which may be involved in accidents or on which accidents may occur. The safety property of a unit in a specified period of time, μ, is the number of accidents by type and severity, expected to occur on or to it. The μ of a unit is determined by its safety-related traits. Units that share some traits form a population. To describe the safety of a population of units the $E\{\mu\}$ and the $\sigma\{\mu\}$ will be used. An SPF is a tool which for a multitude of populations provides estimates $E\{\mu\}$, and of $\sigma\{\mu\}$. The use of these in sketch applications was illustrated.

An SPF is usually an equation on the left-hand side of which are estimates of $E\{\mu\}$ or $\sigma\{\mu\}$ and on the right-hand-side is a function of traits and parameters. From the application-centered perspective the focus is on the left-hand side; from the cause-effect perspective the main interest is in the parameters of the right-hand-side. Throughout this book the application-centered view of SPFs will be the guiding principle.

References

The American Association of State Highway and Transportation Officials (AASHTO) (2010) Highway safety manual, 1st edn. AASHTO, Washington, DC

Barber K (1999) Toronto Star, July 31

Cobb PW (1940) The limit of usefulness of accident rate as a measure of accident proneness. J Appl Psychol 24:154–159

Fisher RA (1922) On the mathematical foundations of theoretical statistics. Philos Transact A Math Phys Eng Sci 222:309–368

Hauer E (1997) Observational before-after studies in road safety. Pergamon, Oxford

Hauer E, Quaye K, Liu Z (1993) On the use of accident or conviction counts to trigger action. Transportation research record 1401. Washington, Transportation Research Board, pp 17–25

Lehmann ER (1990) Model specification: the views of Fisher and Neyman, and later developments. Statist Sci 5(2):160–168

Whitehead AN (1958) An introduction to mathematics. Oxford University Press, New York

A Safety Performance Function for Real Populations

2

Abstract
An SPF provides estimates of the mean and standard deviation of the μ's for many populations of units. When the units of these populations are real, the estimation of their $E\{\mu\}$ and $\sigma\{\mu\}$ is straightforward and their meaning is clear. Attaining this clarity is the main aim of this chapter. A simple SPF for real units will be built using data for 2,228 Colorado road segments.

2.1 The Origin

Chapter 1 introduced the notion that units which share some safety-related traits form a population and that the safety of populations can be parsimoniously and usefully described by the mean of the μ's of its units ($E\{\mu\}$) and by their standard deviation ($\sigma\{\mu\}$). The SPF was said to be a tool which provides estimates of $E\{\mu\}$ and $\sigma\{\mu\}$ for a multitude of populations.

When the term "Safety Performance Function" (SPF) was first coined,[1] it referred to a relationship that gives the average number of accidents for various amounts of exposure.[2] Over the years, the term has been broadened in two directions. First, nowadays the SPF is a function not only of exposure but also of

[1] Hauer (1995).

[2] Exposure is a measure of opportunities for accidents to occur. The most commonly used measure of exposure is "vehicle miles of travel" (VMT). The concept of exposure is tied to that of risk. Risk is usually construed as the probability of a crash of a specified type and severity to occur per unit of exposure which, in probability theory, corresponds to a "trial" the outcome of which is either "accident" or "no accident." These definitions of exposure and risk are due to Hauer (1982). For examples of usage see, e.g., Keall and Frith (1999) and Hakkert et al. (2002).

E. Hauer, *The Art of Regression Modeling in Road Safety*,
DOI 10.1007/978-3-319-12529-9_2

other traits.[3] Second, the SPF now provides estimates not only of the average number of accidents but also of the diversity of the μ's in a population.

The "F" in SPF stands for "function." While "function" evokes the image of an equation, "it ain't necessarily so." A function can also be an algorithm, a graph, or a table; any device which for specific values of "predictor variables" returns a value of the "dependent variable." To secure a clear understanding of the SPF and of the estimates which it provides, it is best to present this first SPF in the form of a table. Doing so will allow one to speak about real populations of units the μ's of which have a real mean $E\{\mu\}$ and a real standard deviation $\sigma\{\mu\}$.

2.2 The Estimate of $E\{\mu\}$

The data to be used throughout the book are described in Sect. 3.2. The subset used here is of 2,228 rural two-lane road segments in Colorado which are between 0.5 and 1.5 miles long.[4] These road segments were sorted into 20 bins by average AADT as shown in Table 2.1.

The ratio of the entries in columns 2 and 3 is in column 4. This is the estimate of $E\{\mu\}$, the primary element of a simple tabular SPF. The squares in Fig. 2.1 are the graphical representation of these $\hat{E}\{\mu\}$.

The ordinate of the squares – the average number of accidents per road segment – is the estimate of $E\{\mu\}$ for real populations. Thus, for example, for the population of road segments defined by the traits (1) State: Colorado, (2) Road Type: two-lane, (3) Setting: rural, (4) Segment Length: 0.5–1.5 miles, and by (5) AADT: 9,000–10,000 vehicles per day, the average number of accidents in 5 years was 5.37(=102/19). This is an estimate of the average of 19 μ's, the $\hat{E}\{\mu\}$ of this population. An SPF of this type makes it clear that what is listed in a row of the table and shown in the graph always pertains to a population of units. Here there are 20 such populations each defined by traits (1)–(4) and by an AADT bin.

The accuracy of this estimate is represented by the standard error of $\hat{E}\{\mu\}$ and shown in Fig. 2.1 by the horizontal bars that surround the squares. To illustrate, for segments with 9,000 < AADT < 10,000, the standard error of $\hat{E}\{\mu\}$ is $\sqrt{102}/19 = \pm 0.53$ accidents in 5 years.[5] Accordingly, the corresponding horizontal bars are placed at 5.37 + 0.53 and at 5.37–0.53.

[3] The Highway Safety Manual (AASHTO 2010, page G-13) defines SPF as "... an equation used to estimate or predict the expected average crash frequency per year at a location as a function of traffic volume and in some cases roadway or intersection characteristics (e.g., number of lanes, traffic control, or type of median)."

[4] To download the data, go to http://extras.springer.com/ and enter the ISBN of this book. The ISBN (International Standard Book Number) is found just after the title page. Look in the "Data" folder for "4 (a or b) Colorado condensed (xls or xlsx)." To make Table 2.1 out the data, the Pivot Table tool described in Sect. 3.3 was used.

[5] The standard error is the estimate of a standard deviation and is usually denoted by s The estimate of the mean of μ's $\equiv \hat{E}\{\mu\}$ = (Number of accidents)/(Number of segments). Assuming that the number of accidents is Poisson distributed, the standard error of the average crash rate is $s = \sqrt{\text{Number of accidents}}/(\text{Number of segments})$.

Table 2.1 A tabular SPF for 0.5–1.5 mile long two-lane rural road segments in Colorado

Data			Estimates		
1	2	3	4	5	6
Average AADT	Injury and fatal accidents 1994–1998	Number of segments	$\hat{E}\{\mu\}$, 1994–1998 I&F accidents/segment	$\hat{\sigma}\{\hat{E}\{\mu\}\}$, standard error of $\hat{E}\{\mu\}$	S^2, sample variance
500	376	975	0.39	0.02	0.53
1,500	445	466	0.95	0.05	1.36
2,500	382	260	1.47	0.08	2.80
3,500	381	178	2.14	0.11	4.47
...	...	...	...	...	...
9,500	102	19	5.37	0.53	35.18
10,500	81	18	4.50	0.50	9.81
...	...	...	...	...	...
18,500	14	1	14.00	3.74	0.00
19,500	13	1	13.00	3.61	0.00
Sum	3,011	2,228			

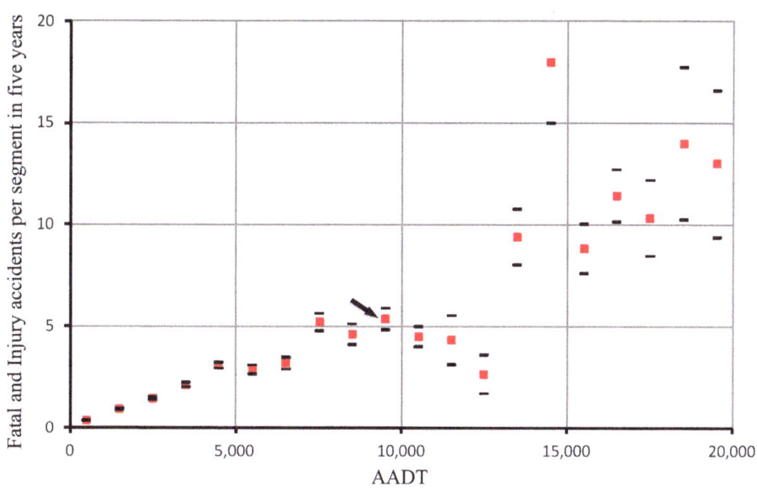

Fig. 2.1 An SPF which is a function but not an equation

The more segments there are in a bin, the more accurately can its $E\{\mu\}$ be estimated.[6] As is evident in Table 2.1, the larger the average AADT the fewer are

[6] The numerator in $s = \sqrt{\text{Number of accidents}}/(\text{Number of segments})$ can be written as $\sqrt{\text{Accidents per segment} \times \text{Number of road segments}}$. It follows that for any given crash rate (≡Accidents/segment) the standard error of $\hat{E}\{\mu\}$ is inversely proportional to the square root of the number of road segments which serve for estimation.

the segments in a bin. This is why, going from left to right in Fig. 2.1, the bars move further and further away from the squares.

What can be said on this basis? One can say that on Colorado rural two-lane road segments which are 0.5–1.5 miles long and have an AADT between 9,000 and 10,000 vehicles, the average of the μ's in the 1994–1998 period was about 5.37 I&F accidents and that the standard error of this estimate is ± 0.53.[7]

Because the $\hat{E}\{\mu\}$ of a population is a clue about the μ's of all units in that population one can say a bit more. Suppose that all we know about unit i is that it is a segment of a rural two-lane road in Colorado, that it is 1.0 miles long, and that it has an AADT of 9,500 vehicles per day. The estimate of $\hat{E}\{\mu\}$ for the population of road segments defined by these traits is 5.37 I&F accidents in 5 years. Because we know nothing about unit i to distinguish it from other segments in that population, 5.37 is also the best estimate of the μ of unit i, the μ_i. It is the "best estimate" because if $\hat{E}\{\mu\}$ was always used to estimate the μ of specific units with traits that match those of the population to which $\hat{E}\{\mu\}$ pertains, the variance of the estimates would be the smallest.

This section was about the estimates of $E\{\mu\}$ for 20 populations that comprise real units. Taken together, these estimates together with their standard error make up the principal element of the SPF. However, for most applications estimates of $E\{\mu\}$ are insufficient. As noted earlier,[8] to determine whether a unit is a "blackspot," to predict what might be the safety benefit of a treatment or of a design change, and to estimate what the safety effect of an intervention was, one needs to know how diverse are the μ's in the population. For these applications, estimates of $\sigma\{\mu\}$ are also needed.[9] This element of the SPF is discussed next.

2.3 The Estimate of $\sigma\{\mu\}$

The $\sigma\{\mu\}$ characterizes the diversity of μ's in a population of units. To explain how an estimate of $\sigma\{\mu\}$ can be extracted from accident counts consider Fig. 2.2.[10]

The circles of the top tier represent the μ's of five units and the squares of the bottom tier represent their accident counts. Obviously there is a relationship

[7] An approximate rule of thumb is that the "true value" is within ± 2 standard deviations of the estimated value 19 times out of 20. This rule is based on the assumption that the estimate is unbiased and normally distributed.

[8] See Sects. 1.3 and 1.4.

[9] The $\hat{\sigma}\{\hat{E}\{\mu\}\}$ in column five of Table 2.1 and the $\sigma\{\mu\}$ to be discussed next are two entirely different constructs. One measures the accuracy with which $E\{\mu\}$ is estimated, the other measures the diversity of μ's in a population. How they combine to determine the accuracy of an estimate of μ for a specific unit is discussed in Sect. 2.4.

[10] Figure 2.2 is the same as Fig. 1.3 and is reproduced here for convenience.

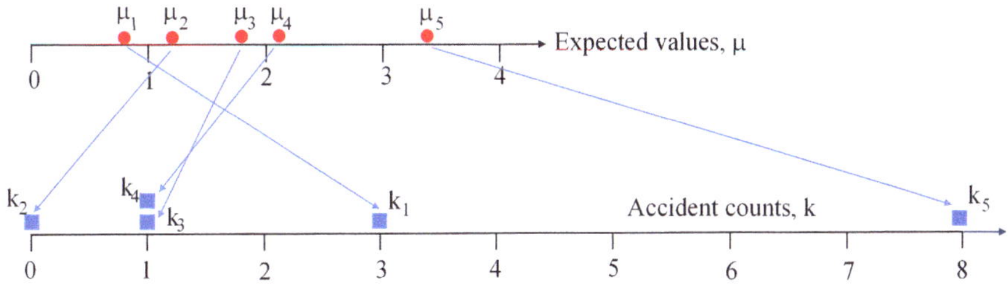

Fig. 2.2 The relationship between μ's and k's

between the diversity of the μ's and that of the k's. It can be shown[11] that when the accident counts for each unit are Poisson distributed then

$$V\{\mu\} = V\{k\} - E\{\mu\}$$

and therefore

$$\hat{\sigma}^2\{\mu\} = \big(\text{Sample variance of accident counts in population} - \text{Sample mean of accident counts in population}\big) \text{ if positive, and 0 otherwise.} \tag{2.1}$$

To illustrate, for the 19 segments with 9,000 < AADT < 10,000 that are 0.5–1.5 miles long the sample mean is 5.37 ± 0.53 accidents in 5 years and the sample variance (Column 6 in Table 2.1) is 35.18 accidents^2. Therefore, $\hat{\sigma}\{\mu\} = \pm\sqrt{35.18 - 5.37} = \pm 5.46$ accidents in 5 years.[12] Proceeding in the same manner for all bins in Table 2.1 for which there is sufficient information[13] Fig. 2.3 shows how in our data $\hat{\sigma}\{\mu\}$ depends on AADT. With this, the second element of this simple SPF, the estimate of $\sigma\{\mu\}$, is also in hand.

2.4 The Two σ's; Homogeneity Versus Accuracy

Two different σ's were discussed in this chapter: $\sigma\{\mu\}$ and $\sigma\{\hat{E}\{\mu\}\}$. The first describes the diversity of μ's amongst the units of a population; the second characterizes the accuracy with which the mean of these μ's is estimated. The nature of both, as well as the differences between them, can be clarified with the help Fig. 2.4. In this figure what is computed from data is in black and what is unknown and estimated is in grey.

[11] For proof, see Appendix C or Hauer (1997), pages 204–205.

[12] If a yearly crash rate is of interest, these results have to be divided by 5.

[13] For AADTs > 11,000, there are too few segments per bin to compute useful estimates of the sample variance of accident counts.

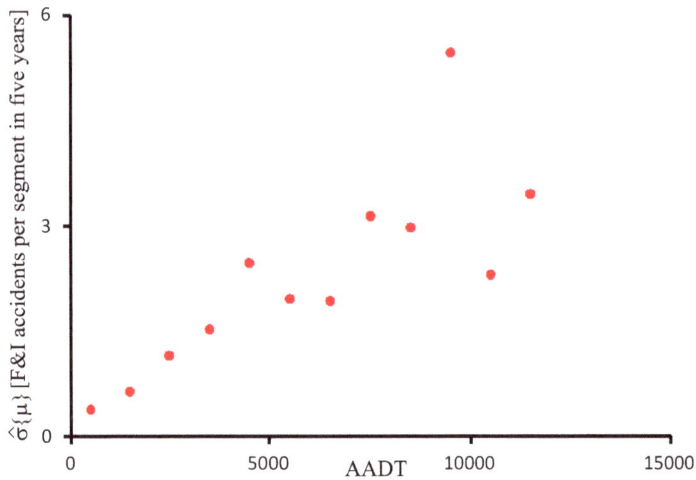

Fig. 2.3 $\hat{\sigma}\{\mu\}$, the second element of an SPF

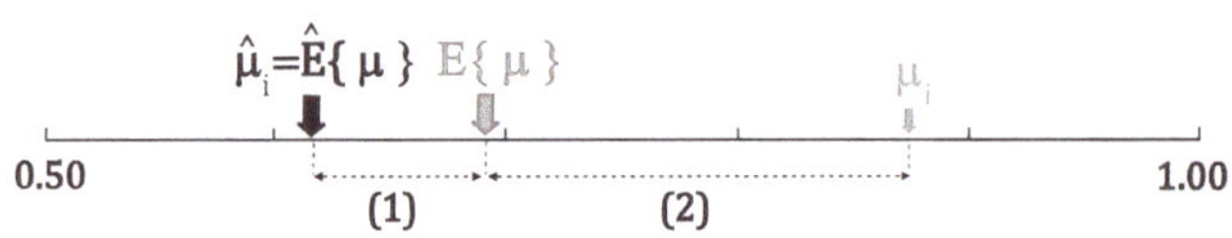

Fig. 2.4 Two variance components

The expected value of the square of distance (1) measures the accuracy[14] with which $E\{\mu\}$ is estimated; it is the $\sigma^2\{\hat{E}\{\mu\}\}$. The expected value of the square of distance (2) measures how diverse are the μ's of the units in population; it is the $\sigma^2\{\mu\}$. For the real populations and data in Table 2.1, estimates of both σ's are shown in Fig. 2.5.

In important applications both σ's are in play. Thus, for example, suppose that we are interested in the accuracy with which μ_i (the μ of unit i) is estimated when $\hat{E}\{\mu\}$ serves as its estimator[15]. That is, we need the variance of the sum of distances (1) and (2) in Fig. 2.4. In principle, the accuracy with which $E\{\mu\}$ is estimated and the diversity of the μ's in a population are causally independent.[16] If so,

---

[14] According to the Joint Committee for Guides in Metrology (JCGM 2008), the term "accuracy" refers to the degree of closeness of measurements of a quantity to that quantity's true value. In contrast, the word "precision" refers to the degree to which repeated measurements under unchanged conditions show the same results. To illustrate, if an experiment contains a systematic error then repeating the same flawed experiment would yield a string of possibly precise but still inaccurate (biased) results. Eliminating the systematic error would improve accuracy but may not change precision of the results.

[15] An estimator is a rule for calculating an estimate from data. That $\hat{E}\{\mu\}$ is the estimator of $\hat{\mu}_i$ is indicated in Fig. 2.4 by setting $\hat{\mu}_i$ to equal $\hat{E}\{\mu\}$.

[16] In estimation, however, the two summands must be correlated inasmuch as the statistic $\sqrt{\text{Number of accidents}}/(\text{Number of segments})$ features in both.

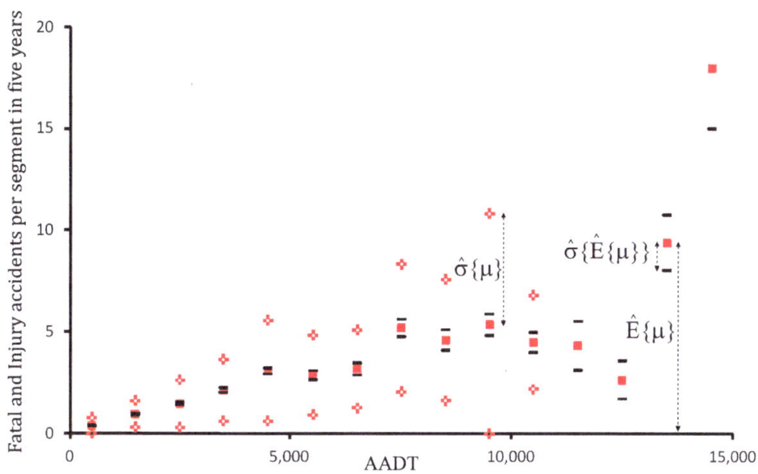

Fig. 2.5 Estimates of $\sigma\{\mu\}$ versus $\sigma\{\hat{E}\{\mu\}\}$

$$\sigma\{\hat{\mu}_i\} = \sqrt{\sigma^2\{\hat{E}\{\mu\}\} + \sigma^2\{\mu\}} \tag{2.2}$$

To illustrate, consider again that segment of a rural two-lane road in Colorado that is 1.0 mile long and has an AADT of 9,500 vehicles. If to estimate its μ we use the $\hat{E}\{\mu\} = 5.37$ then, by Eq. 2.2, $\sigma\{\hat{\mu}_i\} = \sqrt{0.53^2 + 5.46^2} = \pm 5.47$ accidents in 5 years.[17]

Equation 2.2 leads to a key insight. When dealing with real populations (such as that of the Colorado road segments), the $\hat{E}\{\mu\}$ is obtained by dividing the number of accidents in a bin by the number of segments in it. The wider the bin the more segments will be in it and therefore the smaller will be the $\sigma\{\hat{E}\{\mu\}\}$. However, the wider the bin the more diverse will tend to be the μ's of the units in it. If so, making the bin wider will usually cause $\sigma\{\mu\}$ to increase. The compromise between these opposing tendencies defines a valley in the sum of the radical in Eq. 2.2 and thereby an optimal bin width.

The same key insight and same conflicting tendencies will apply when, instead of tabulating bin averages, the SPF will be obtained by fitting a function to data. In this case it will not be the bin width that matters but the number of variables in the function. The larger the number of variables the larger will tend to be the uncertainty surrounding the estimates of $E\{\mu\}$ but the less diverse will be the μ's of units belonging to the imagined populations.

[17] In this case, the accuracy with which the μ can be estimated is governed by the diversity of μ's in the population of units with the same traits, and not by the accuracy by which the mean of the μ's is estimated.

2.5 Summary

An SPF consists of two main elements: (1) Estimates of $E\{\mu\}$, the mean of the μ's in each population and the standard deviation of this estimate, the $\sigma\{\hat{E}\{\mu\}\}$; (2) estimates of $\sigma\{\mu\}$, the standard deviation of the μ's in each population. Both are needed for practical applications. In this chapter a simple table-and-graph SPF was built out of data for real units. The aim was to make the SPF tangible and the meaning of $E\{\mu\}$ and $\sigma\{\mu\}$ unambiguous. It is now clear that whatever is measured on the vertical axes of Figs. 2.1, 2.3, or 2.5 pertain to populations of units.

When dealing with real populations consisting of many units, the estimation of $E\{\mu\}$ and $\sigma\{\hat{E}\{\mu\}\}$ is straightforward. To estimate $\sigma\{\mu\}$ Eq. 2.1 can be used. Equation 2.2 can be used to estimate the standard error of $\hat{\mu}_i$ when $\hat{E}\{\mu\}$ is its estimator.

The simple SPF used in this section was sufficient to illustrate its essential nature and to highlight its two main elements. However, to be of practical use, the simple tabular SPF has to be enriched along three lines. First, the bins are too broad. Segment Length is usually known to the next hundredth of a mile and it does not make sense to use bins in which the length of segments may vary by as much as a mile. Second, while the simple SPF in this chapter accounts for Segment Length and AADT, many safety-related traits (variables) are missing. Thus, example, one cannot know what is normal for a road segment without taking into account its lane width. Neither can one meaningfully compare the mean of μ's for two populations without considering possible differences in, say, terrain. Nor does it make sense to use the mean of the μ's to estimate the μ of unit i if it is on a sharp curve while the curvature of the segments in the population is unknown. For the SPF of practical use, one has to add to the SPF some important population-defining traits. Third, one may expect that underneath the squares in Fig. 2.1 or the circles in Fig. 2.3 there is some presently unknown continuous curve. After all, AADT and Segment Length are nearly continuous variables. Which traits to add, what tools to use for this purpose, and how to fit mathematical functions to data are the question discussed in detail in later chapters.

References

AASHTO (The American Association of State Highway and Transportation Officials) (2010) Highway safety manual, 1st edn. Washington, DC, AASHTO

Hakkert AS, Braimaister L, van Schagen I (2002) The uses of exposure and risk in road safety studies. In: Proceedings of the European Transport Conference. Homerton College, Cambridge

Hauer E (1982) Traffic conflicts and exposure. Accid Anal Prev 14(5):359–364

Hauer E (1995) On exposure and accident rate. Traffic Eng Contr 36(3):134–138

Hauer E (1997) Observational before-after studies in road safety. Pergamon, Oxford

JCGM (Joint Committee for Guides in Metrology) (2008) International vocabulary of metrology – basic and general concepts and associated terms. 200:2008

Keall MD, Frith WJ (1999) Measures of exposure to risk of road crashes in New Zealand. IPENZ Transactions 26(1/CIV):7–12

Exploratory Data Analysis

3

Abstract

The role of an exploratory data analysis (EDA) is to equip the modeler with an understanding of the data. More specifically, an EDA helps to answer two core questions: (a) whether a trait is safety related and (b) what function can be used to represent it in the model equation. This chapter shows how to do an EDA of the Colorado data using spreadsheet tools. As expected, Segment Length, AADT, and Terrain are safety-related traits. However, one cannot say what function links the $E\{\mu\}$ and these variables. The numerical results of the EDA motivate important general observations.

3.1 Introduction

As noted earlier this book has two objectives: (a) to teach how to fit a multivariable statistical model to data using a simple spreadsheet and (b) to promote the understanding that is at the core of good modeling. To set the stage the preceding two chapters were devoted to objective (b). This chapter is addressing both objectives.

In line with the first objective, it will teach how to conduct an exploratory data analysis (EDA) of the Colorado data using the Excel spreadsheet tools. The assumption is that the reader has working knowledge of Excel but may not be familiar with the Pivot Table tool. Because the Pivot Table plays a central role in the EDA its use will be explained in detail.

In line with the second objective, EDA is about insight into the message of the data. Models are built from data. The transformation of data into a model is the work of a modeler. The modeler has to make a variety of choices: what traits (variables) to use to define the populations, what functions to use for combining the variables into a model equation, what should be minimized of maximized to estimate parameter values and to obtain a good fit, how can the fit be improved, which data points are outliers, etc. These choices depend on the exercise of

E. Hauer, *The Art of Regression Modeling in Road Safety*,
DOI 10.1007/978-3-319-12529-9_3

judgment and insight. This chapter is about developing an initial insight into what the data suggest.

"Some parts of EDA are ugly, but the real world is ugly, particularly when errors and other aberrant material enter a data set." Brillinger (2002) writing about the pioneer of EDA, J.W. Tukey

This familiarization exercise is called "Initial Exploratory Data Analysis[1]." The tools of EDA are many and so are the corresponding specialized software packages. But EDA is more than a collection of tools; it is an approach to understanding the message of data.[2]

The EDA can help with answering two core questions:

(a) Whether a trait (variable) has an orderly relationship[3] with the $E\{\mu\}$.
(b) If the relationship is orderly, what function can represent it.

The same two questions will be asked repeatedly whenever the introduction of a new variable into the model is considered. In this case a special-purpose Variable Introduction EDA (the VIEDA) will be used.[4]

The EDA produces tables and graphs the purpose of which is to bring about an understanding of and intimacy with the data from which a model is to be built. While the numerical results of the EDA pertain to specific data, they will give rise to some general observations. These general observations have practical consequences.

---

[1] EDA is usually thought of as a set of activities that precedes formal modeling and, in this sense, calling it "initial" may seem redundant. However, as will be stressed repeatedly, modeling is not a once-through process. Rather, it is akin to a spiral the coils of which are cyclically repeated activities. The EDA will be an integral part of every modeling cycle, every turn of the spiral. It will help to determine whether the data indicate that a new trait is to be added to the SPF and in what form and way. In this role it will be called a Variable Introduction EDA or, by acronym, a VIEDA.

[2] Here is what the Engineering Statistics Handbook (NIST/SEMATEC) says: "EDA is an approach to data analysis that postpones the usual assumptions about what kind of model the data follow with the more direct approach of allowing the data itself to reveal its underlying structure and model. EDA is not a mere collection of techniques; EDA is a philosophy as to how we dissect a data set; what we look for; how we look; and how we interpret." (Sect. 1.1).

[3] An orderly relationship is one where the existence of a perceived pattern makes curve-fitting a sensible choice. If the relationship is not orderly there is no reason to add this trait to the SPF. The question of when a trait should be added to the SPF is discussed at length in Chap. 9.

[4] The VIEDA is discussed in Sect. 9.2.

3.2 The Data

The data used for the EDA and for all subsequent illustrations are for two-lane rural roads in Colorado.[5] They pertains to 5,323 road segments of varying length totaling 6,029 miles, with an AADT up to about 20,000 (average 2,151), on which, during the 13 years between 1986 and 1998, 21,718 injury and fatal and 52,317 total accidents were recorded. The data for the first 9 (of 5,323) road segments are shown in Fig. 3.1.

The main feature of data of this kind is that they describe a collection of units without there being a specific intervention or treatment applied at a certain time. Such data are usually called "cross-sectional."

Colorado, two-lane, rural

	A	B	C	D	H	I	J	N	O	P	Q
2		Length	ADT							Injury + Fatal	
3	Segment No.	Miles	1986	1987	1991	1992	1993	1997	1998	1986	1987
4	1	1.50	60	60	60	63	50	64	66	0	0
5	2	2.90	60	60	110	116	100	138	144	0	0
6	3	1.49	60	60	110	116	100	138	144	0	0
7	4	0.39	60	60	110	116	100	138	144	0	0
8	5	3.89	60	60	110	116	100	138	144	0	0
9	6	2.03	100	100	110	116	100	138	144	0	0
10	7	0.91	100	100	100	105	100	117	122	0	0
11	8	0.89	0	0	100	104	110	147	153	0	0
12	9	0.97	100	100	120	116	140	159	165	0	1

AA	AB	AC	AD	AE	AF	AO	AP	AQ	AR
			Total Accidents						Terrain
1997	1998	Sum	1986	1987	1988	1997	1998	Sum	F,R or M
0	0	0	1	0	0	0	0	1	F
0	0	0	0	0	0	0	0	1	F
0	0	0	0	0	0	0	0	1	F
0	0	0	0	0	0	0	0	0	F
0	0	0	0	0	0	0	0	1	F
0	0	0	1	0	0	1	0	2	F
0	0	0	0	0	0	0	0	0	F
0	0	0	0	0	0	0	0	0	F
0	0	1	0	1	0	0	0	1	F

Fig. 3.1 A few rows of the Colorado data

[5] The data can be downloaded from http://extras.springer.com/ using the ISBN (International Standard Book Number) of this book. Look in the "Data" folder for file "1 (a or b) Colorado full. (xls or xlsx)."

Data come with holes and errors. Some such data problems can be detected early. The spreadsheet environment is convenient for this purpose; holes are visible, inconsistencies are easy to check for. Early detection of missing and erroneous data will save the pain of having to redo what has already been thought completed.

Some data errors become manifest in the form of "outliers." An outlier is an observation that appears to be unlikely in view of most other observations. The problem is that what may look to be an outlier in the early stages of modeling may be seen as quite normal when all variables are in the model. This is why, inconveniently, outliers cannot be dealt with early on.

In the Colorado data the holes were already plugged and obvious inconsistencies corrected. However, some holes cannot be plugged. Thus, for example, in Fig. 3.1 there are no AADT estimates for segment 8 during the early years. Holes which are "missing data" have to be accounted for later during the model fitting stage.

Having introduced the data the EDA work can begin. The aim here is to form an idea about how the average number of accidents varies as a function of Segment Length, AADT, and Terrain. With this aim in mind a condensed data set was prepared.[6] It consists of the Segment Length, the 5-year average AADT for the years 1994–1998, the corresponding 5-year sum of injury and fatal accidents, and Terrain as shown in Fig. 3.2. The condensed data will be used in the EDA and in the early stages of modeling.

	A	B	C	D	E	F
1	CDOT, Rural, two-lane, all terrains.					
2	Segment	Length	Average			
3	Number	Miles	AADT 1994-1998	I&F 1994-1998	Total 1994-1998	Terrain
4	1	1.50	62.2	0	0	F
5	2	2.90	134.8	0	0	F
6	3	1.49	134.8	0	1	F
7	4	0.39	134.8	0	0	F
8	5	3.89	134.8	0	1	F
9	6	2.03	134.8	0	1	F
10	7	0.91	126.2	0	0	F
11	8	0.89	132.0	0	0	F

Fig. 3.2 Condensed data for 1994–1998

[6] To access and download the condensed data go to http://extras.springer.com/ and enter the ISBN of this book. Look in the "data" folder for files "4 (a or b) Colorado condensed. (xls or xlsx)."

3.3 The Pivot Table

To say how $E\{\mu\}$ changes as a function of Segment Length and AADT we need a table with AADT bins as rows, Segment Length bins as columns, and table cells that give statistics such as "number of segments in cell" and "number of accidents in a cell". The Excel spreadsheet has a versatile tool for the job: the "Pivot Table." Readers familiar with the use of the Pivot Table may skip the detailed instructions below.

To introduce the tool click on the "Insert" tab and choose "Pivot Table."[7] The window in Fig. 3.3 opens.

When selecting the data in the "Table/Range" one has to include the headings row (row 3 in Fig. 3.2). When OK is clicked the "Pivot Table Field List" in Fig. 3.4 appears.

To create a table in which the rows will be AADT bins drag the "AADT 1994–1998" field down into the empty square under "Row L. . . ." This opens the "Row Labels" window in Fig. 3.5.

When any number under the "Row Labels" is right-clicked the menu with various actions in Fig. 3.6 opens.

After "Group" is clicked the window in Fig. 3.7 opens.

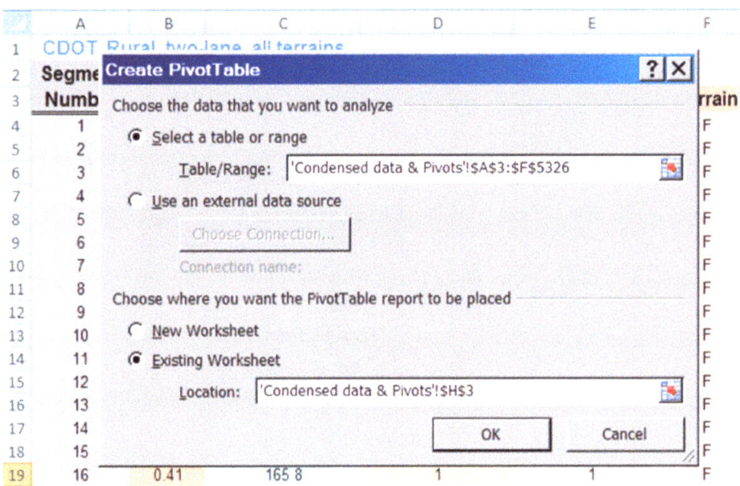

Fig. 3.3 Selecting data and location for pivot table

[7] In all examples I will refer to Excel 2007. Other versions may differ from it in toolbar arrangements, menus and similar inessential detail. Readers may have to adapt to the specifics of the version they use. The first appearance of the Pivot Table in Excel was in version 5 (1993). Up to and including "Office 2013," there are 12 versions of Microsoft Excel.

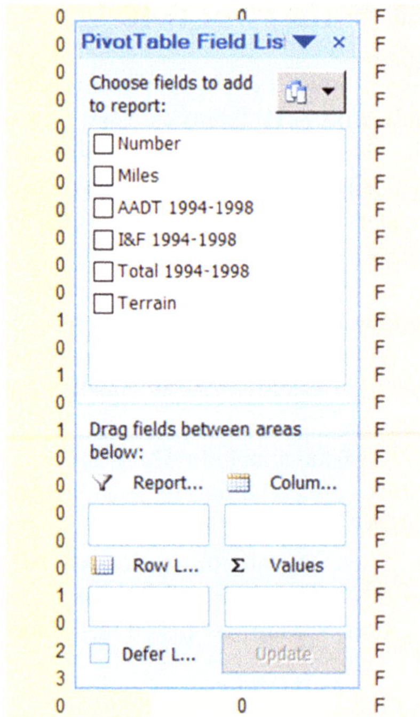

Fig. 3.4 Pivot table fields before selections

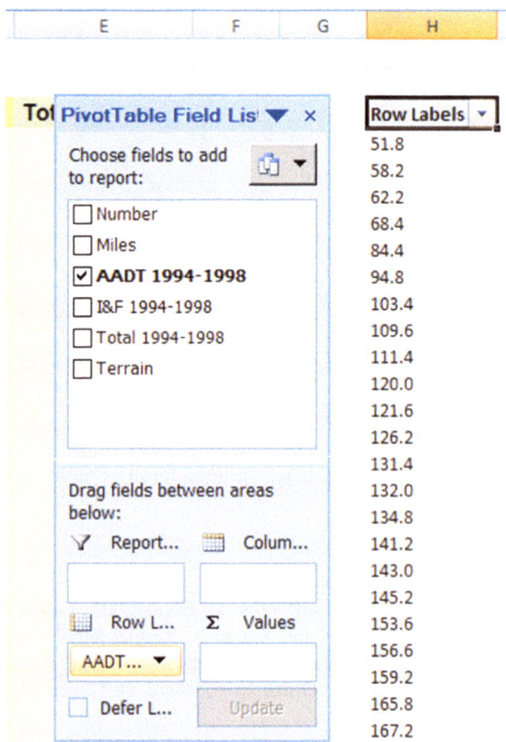

Fig. 3.5 Specifying the "Row Label"

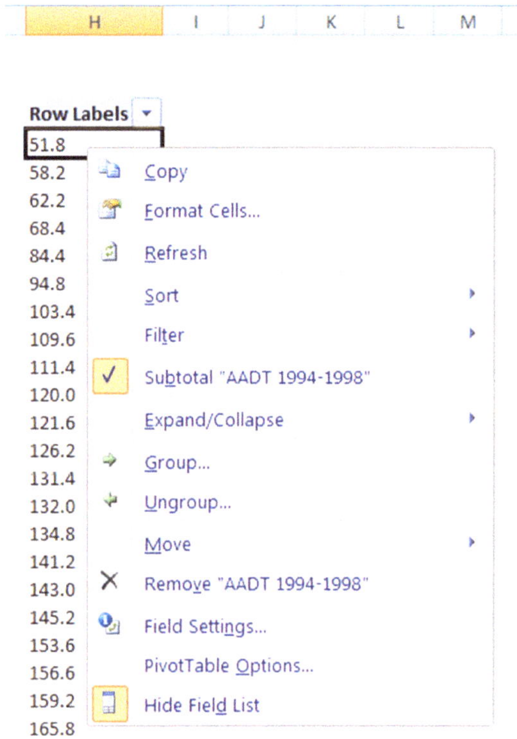

Fig. 3.6 Menu with the "Group" action

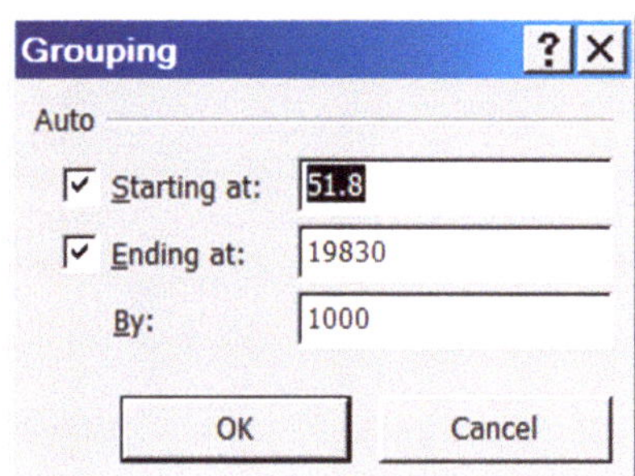

Fig. 3.7 The "Group" window

The automatic entries that appear show the lowest and the highest average AADT as well as a suggested "bin width." Entering 0 instead of 51.8, 20,000 instead of 19,830, keeping 1,000 in the "By" window, and clicking on "OK" leads to Fig. 3.8.

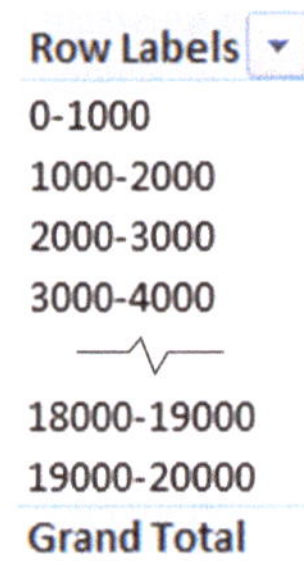

Fig. 3.8 The result of grouping

This completes the preparation of the row bins. The preparation of the column bins is similar. Dragging the "Miles" from the "Pivot Table Field List" into the empty square under "Colum…" opens the window on the right of Fig. 3.9.

Right clicking on any column label (say on 0.01), selecting "Group" on the menu that opens, choosing to start the groups at 0.5 miles, to end at 20 miles, and to make the bins 1 mile wide prepares the table that is now ready to receive a variety of contents.

Dragging the "I&F Sum 1994–1998" field from the "PivotTable Field List" into the empty square under the "Σ Values" heading opens the table in Fig. 3.10.

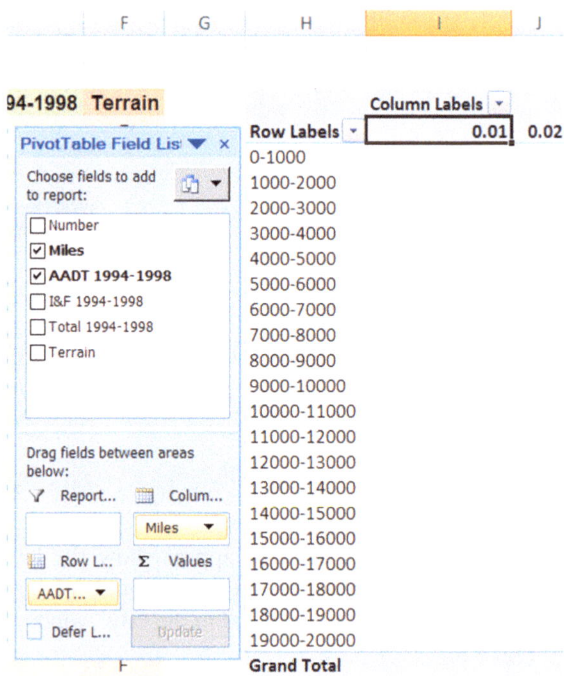

Fig. 3.9 Ungrouped column labels

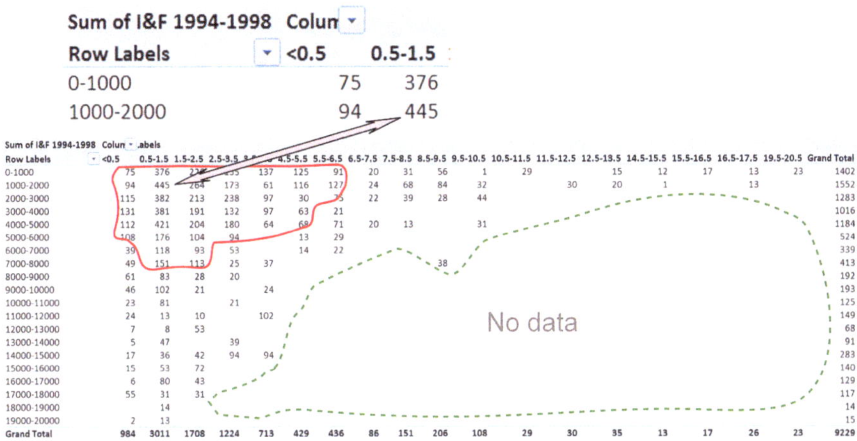

Row Labels	<0.5	0.5-1.5	1.5-2.5	2.5-3.5	3.5-4.5	4.5-5.5	5.5-6.5	6.5-7.5	7.5-8.5	8.5-9.5	9.5-10.5	10.5-11.5	11.5-12.5	12.5-13.5	14.5-15.5	15.5-16.5	16.5-17.5	19.5-20.5	Grand Total
0-1000	75	376	226	155	137	125	91	20	31	56	1	29		15	12	17	13	23	1402
1000-2000	94	445	264	173	61	116	127	24	68	84	32		30	20	1		13		1552
2000-3000	115	382	213	238	97	30	75	22	39	28	44								1283
3000-4000	131	381	191	132	97	63	21												1016
4000-5000	112	421	204	180	64	68	71	20	13		31								1184
5000-6000	108	176	104	94		13	29												524
6000-7000	39	118	93	53		14	22												339
7000-8000	49	151	113	25	37					38									413
8000-9000	61	83	28	20															192
9000-10000	46	102	21		24														193
10000-11000	23	81		21															125
11000-12000	24	13	10		102														149
12000-13000	7	8	53																68
13000-14000	5	47		39															91
14000-15000	17	36	42	94	94														283
15000-16000	15	53	72																140
16000-17000	6	80	43																129
17000-18000	55	31	31																117
18000-19000		14																	14
19000-20000	2	13																	15
Grand Total	984	3011	1708	1224	713	429	436	86	151	206	108	29	30	35	13	17	26	23	9229

Fig. 3.10 Injury and fatal accident counts

It shows the number of I&F accidents in each bin. Thus, for example, there are 445 accidents on 0.5–1.5 miles long segments with $1{,}000 < \text{AADT} < 2{,}000$ vehicles per day. The solid line is the boundary of the region where somewhat reliable guidance may be found; the dashed line surrounds the region with no data.

To summarize the data differently right click anywhere within the table and the menu in Fig. 3.11 opens. Choosing "Count" would produce a table with number of segments in each bin; choosing "Average" yields the table of "average accidents/ segment" $\left(=\hat{E}\{\mu\}\right)$ a part of which is shown in Fig. 3.12.

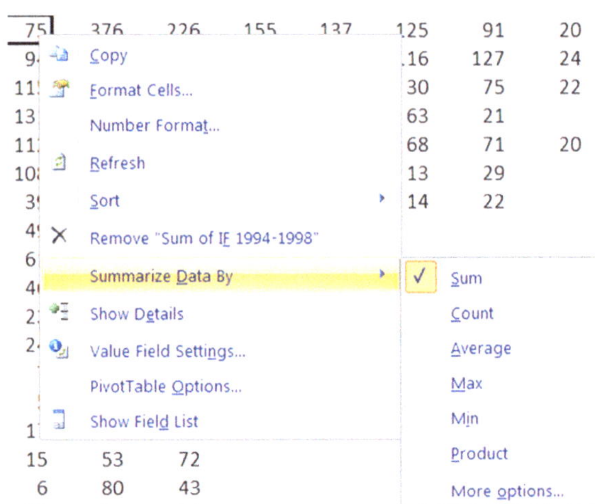

Fig. 3.11 Choosing the content of the table

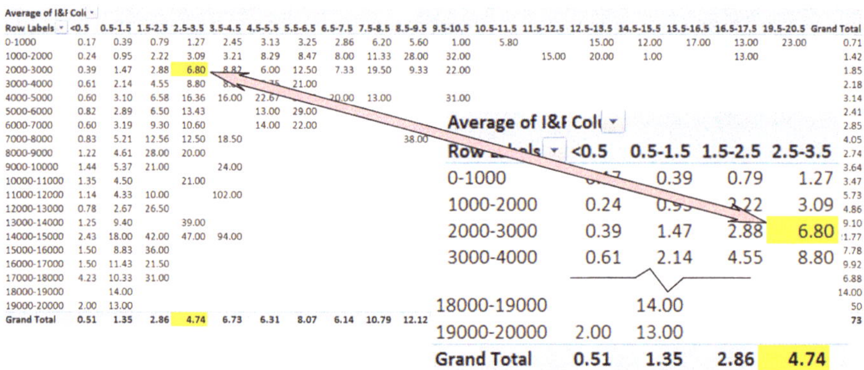

Average of I&F Coll

Row Labels	<0.5	0.5-1.5	1.5-2.5	2.5-3.5	3.5-4.5	4.5-5.5	5.5-6.5	6.5-7.5	7.5-8.5	8.5-9.5
0-1000	0.17	0.39	0.79	1.27	2.45	3.13	3.25	2.86	6.20	5.60
1000-2000	0.24	0.95	2.22	3.09	3.21	8.29	8.47	8.00	11.33	28.00
2000-3000	0.39	1.47	2.88	6.80	8.82	6.00	12.50	7.33	19.50	9.33
3000-4000	0.61	2.14	4.55	8.80	[illegible]	[illegible]	21.00			
4000-5000	0.60	3.10	6.58	16.36	16.00	22.67	[illegible]	20.00	13.00	
5000-6000	0.82	2.89	6.50	13.43		13.00	29.00			
6000-7000	0.60	3.19	9.30	10.60		14.00	22.00			
7000-8000	0.83	5.21	12.56	12.50	18.50					38.00
8000-9000	1.22	4.61	28.00	20.00						
9000-10000	1.44	5.37	21.00		24.00					
10000-11000	1.35	4.50		21.00						
11000-12000	1.14	4.33	10.00		102.00					
12000-13000	0.78	2.67	26.50							
13000-14000	1.25	9.40		39.00						
14000-15000	2.43	18.00	42.00	47.00	94.00					
15000-16000	1.50	8.83	36.00							
16000-17000	1.50	11.43	21.50							
17000-18000	4.23	10.33	31.00							
18000-19000		14.00								
19000-20000	2.00	13.00								
Grand Total	**0.51**	**1.35**	**2.86**	**4.74**	**6.73**	**6.31**	**8.07**	**6.14**	**10.79**	**12.12**

Row Labels	9.5-10.5	10.5-11.5	11.5-12.5	12.5-13.5	14.5-15.5	15.5-16.5	16.5-17.5	19.5-20.5	Grand Total
0-1000	1.00	5.80		15.00	12.00	17.00	13.00	23.00	0.71
1000-2000	32.00		15.00	20.00	1.00		13.00		1.42
2000-3000	22.00								1.85
3000-4000									2.18
4000-5000	31.00								3.14
5000-6000									2.41
6000-7000									2.85
7000-8000									4.05
8000-9000									2.74
9000-10000									3.64
10000-11000									3.47
11000-12000									5.73
12000-13000									4.86
13000-14000									9.10
14000-15000									1.77
15000-16000									7.78
16000-17000									9.92
17000-18000									6.88
18000-19000									14.00
19000-20000									50
Grand Total									**73**

Average of I&F Coll

Row Labels	<0.5	0.5-1.5	1.5-2.5	2.5-3.5
0-1000	[illegible]	0.39	0.79	1.27
1000-2000	0.24	[illegible]	[illegible]	3.09
2000-3000	0.39	1.47	2.88	6.80
3000-4000	0.61	2.14	4.55	8.80
18000-19000		14.00		
19000-20000	2.00	13.00		
Grand Total	**0.51**	**1.35**	**2.86**	**4.74**

Fig. 3.12 I&F accidents in 5 years per segment

The question was how $E\{\mu\}$ changes as a function of Segment Length and AADT. Inasmuch as the entries of the table in Fig. 3.12 are estimates of $E\{\mu\}$ the question has now been answered. Pictures are easier to interpret than tables of numbers and therefore the next step is to present the numerical results of Fig. 3.12 as graphs. However, even before proceeding with the visualization task, the numerical results already obtained call for some important observations to be made.[8] At this point the discussion of EDA is put temporarily on hold in order to point out what is perhaps obvious but could be easily missed.

3.4 Pausing for Reflection

Suppose that we are interested in the μ of a road segment about which we know only that it is 3.0 miles long and is a part of the rural two-lane road network in Colorado. Based on the highlighted entry in the last row of Fig. 3.12 the $\hat{E}\{\mu\}$ for such segments is 4.74 F&I accidents in 1994–1998. This is also an unbiased estimate of the μ of the 3.0 miles long segment of interest. If we also knew that the AADT on that segment was 2,500 then, based on the corresponding highlighted cell in Fig. 3.12, $\hat{E}\{\mu\} = 6.80$ F&I accidents in 1994–1998. Now this would be the unbiased estimate of its μ. These two numbers, 4.74 and 6.80, illustrate a general truth:

Obvious Observation 1: Populations defined by different safety-related traits have differing $E\{\mu\}$'s.

[8] Some can reach general conclusions by logical reasoning unaided by numbers and graphs; others need the stimulus and support of numerical results in their quest for reasoned deductions. The EDA suits the latter group.

It follows that when $\hat{E}\{\mu\}$ helps to estimate the μ of a unit of interest[9] then, what we estimate its μ to be depends on which safety-related traits of that unit we have information about. It also follows that:

Obvious Observation 2: For the $\hat{E}\{\mu\}$ *of a population to be an unbiased*[10] *estimator of the* μ *of a unit of interest, the known traits of that unit must be the same as the traits which define the population of units for which* $E\{\mu\}$ *is the mean.*

The variables in the SPF must be the same as the data that are available for the practical task in which the SPF is used.

These "obvious observations" lead to a "not-so-obvious" but important conclusion. Namely, that what traits (variables) are to be used in an SPFs depends on the use for which the SPF is intended and on the information that the users have. To illustrate, if the data available for network screening[11] contain no information about, say, shoulder width or "roadside hazard rating," then shoulder width and roadside hazards rating should not be variables in the SPF used in this activity. For, if they were amongst the SPF variables, it would be unclear what value to use in the SPF to ensure that the estimate of $E\{\mu\}$ is unbiased. However, if in some other circumstance the SPF is to be used for an Empirical Bayes estimate of the μ of a road segment for which, say, the pavement friction and sight distance are known, then the SPF must contain these variables. Otherwise, the estimate of $E\{\mu\}$ which the SPF provides would not be for the population to which that road segment belongs. In sum, one SPF does not fit all practical purposes and many SPFs are needed to match the variety of data available to users.

Obvious Observation 2 puts SPF development on an unaccustomed footing. The usual aim of modelers is to have an SPF that is "well specified."[12] A well specified SPF is made up of all the important safety-related traits.[13] In the pursuit of this aim modelers attempt to assemble data about many traits (variables). After weeding out those variables the parameters of which do not reach the required statistical significance[14] and after eliminating variables the parameters of which are not in

[9] This is the case whenever the μ of a unit is estimated by the Empirical Bayes method.

[10] Bias is the difference between the average value of the estimate and the true value of what is being estimated. If the difference is not zero the estimator is said to be biased.

[11] Network screening is the activity of identifying "blackspots" or "sites with promise," units which may require attention and perhaps remediation The SafetyAnalyst software (Harwood et al. 2010) uses SPFs and Empirical Bayes estimates for network screening.

[12] In regression analysis the process of model specification consists of selecting an appropriate functional form and of choosing the predictor variables. Common errors of model specification are (1) choosing an incorrect functional form, (2) omitting predictor variables which have a relationship with both the dependent variable and one or more of the predictor variables, (3) including an irrelevant predictor variables, and (4) assuming that the predictor variables are measured without error. If an estimated model is misspecified, it will produce biased and inconsistent estimates.

[13] What determines whether a trait is safety-related is discussed in Sects. 9.1 and 9.2.

[14] See e.g., Chiou and Fu (2013, p. 77) and Chen and Persaud (2014, p. 135).

agreement with some prior opinion,[15] modelers tend to report the parameter estimates for the remaining "fully loaded" model. However, when the SPF is to serve for the practical purposes described in Sects. 1.3 and 1.4 the inclusion in the SPF of variables for which data are not routinely available to the user creates a problem. The implications of this shift in aim are several. First, it will affect the decision about whether a variable should be added to the SPF. Second, it will influence the manner in which SPFs are to be reported. Both issues will be discussed in detail later.

The numerical results of the initial EDA allow one to make yet another important "obvious observation."

Obvious Observation 3: *The larger is the number of traits that define a real population, the fewer are the observations from which its* $E\{\mu\}$ *is estimated and the larger tends to be the standard error of the estimate of* $E\{\mu\}$.

To explain, recall that the estimates of $E\{\mu\}$ in Fig. 3.12 were computed by adding up the number of accidents in a bin and dividing by the number of segments in that bin. If the accident counts are Poisson distributed, each such estimate has a standard error of $\pm\sqrt{\text{Number of accidents}}/\text{Number of segments}$. Thus, for example, for the population defined by 2.5 < Segment Length < 3.5 miles the estimate of $E\{\mu\}$ is 1,224/258 = 4.74 accidents in 1994–1998 per segment with a standard error of $\pm\sqrt{1224}/258 = \pm 0.14$. However, if we define a subpopulation by adding the trait that 2,000 < AADT < 3,000, then the estimate of $E\{\mu\}$ is 238/35 = 6.80 accidents per segment and its standard error is $\pm\sqrt{238}/35 = \pm 0.44$.

Subpopulations always contain fewer units than their parent populations. The increase in the standard error is the consequence of creating subpopulations (subsets) by the adding traits and thereby estimating from less data. This then is another consideration for adding traits to the SPF. Not adding a safety-related trait to the SPF will result in a bias according to "Obvious observation 1"; adding the same trait to the SPF will reduce the accuracy with which $E\{\mu\}$ can be estimated. The right course of action is based on balancing these two considerations.

The business of the EDA is to convert data into numbers and then to transform numbers into insight. The insights in this section were at once obvious and unexpected. At this point we return to the EDA and the question of how $E(\mu\}$ depends on Segment Length and AADT.

3.5 Visualization

The purpose of the EDA is to gain insight into the message of the data; this is best done by visualizing it. Three-dimensional graphs may be difficult to interpret. It seems more instructive to reduce graphs to two dimensions, keeping constant all

[15] See e.g., Gross et al. (2013, p. 236) who say that "Additional variables were considered based on available data and included in the models if (1) the variable significantly improved the model, and (2) the effect of the variable was intuitive."

variables but one. Thus, for example, Fig. 3.13 shows how $\hat{E}\{\mu\}$ changes as a function of AADT holding Segment Length approximately constant.[16]

As expected, $\hat{E}\{\mu\}$ increases with AADT. However, even though the data set is rich (6,029 miles of road and 5 years of accident history), one cannot really say whether the relationship is one of proportionality or whether some other mathematical function promises to fit the data well.[17] The randomness of the accident counts, the paucity of data in many cells, and the many unaccounted-for traits which are the source of diversity within each cell, all work to obscure the underlying relationship. Whether some curve-fitting magic can help to reveal that relationship will be examined in the next chapter.

How $\hat{E}\{\mu\}$ varies with Segment Length when AADT is held approximately constant is shown in Fig. 3.14. As expected, $\hat{E}\{\mu\}$ is seen to increase with Segment Length and there is perhaps a hint of nonlinearity.[18]

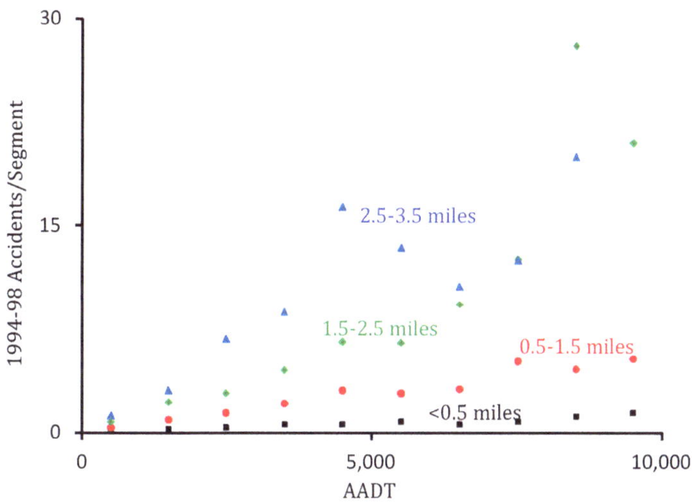

Fig. 3.13 How $\hat{E}\{\mu\}$ changes as a function of AADT within Segment Length bins

[16] The data are from that region of Fig. 3.10 where accident counts are most numerous.

[17] A non-linear relationship with AADT is a common empirical finding. This may reflect the fact that many things change with traffic flow: speed, spacing between vehicles, alertness, etc. In addition many safety-related traits are associated with traffic flow: level of enforcement and maintenance, presence of illumination, road design standards, etc. It would be indeed strange if such a complex interplay of influences when represented by the single trait-AADT, ended up as a straight-line relationship. Even more generally, it would be unexpected for a complex web of causes to become manifest as a simple mathematical function.

[18] One might think that if a 1 mile long segment is expected to have X accidents then a 2 miles long segment with similar traits will have 2X accidents. The problem is that in our data Segment Length may be correlated with various safety-related traits; segment length may have something to do with the way roads are parsed for entry into data bases. Segments tend to end at intersections, jurisdiction boundaries and various geographic features. When intersections are far apart or a

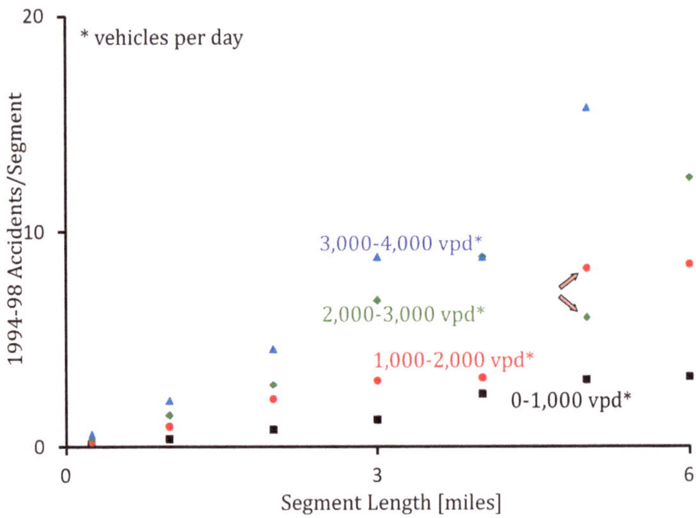

Fig. 3.14 How $\hat{E}\{\mu\}$ changes as a function of Segment Length within AADT bins

Graphs of this kind are a useful hint to how the dependent variable ($E\{\mu\}$ in this case) varies as a function of a predictor variable. However, one must not think that what is seen in a two-dimensional graph represents the workings of a single variable. Causal factors are many and their web is complex. To illustrate, the arrows in the figure point to two estimates of $E\{\mu\}$ the mutual position of which seems anomalous. A reversal of this kind might occur, for example, if a larger proportion of the segments represented by the circle were in mountainous terrain. In this case "Terrain" would be a "lurking variable," a "confounder."[19] In road safety there is never enough observational data to examine how the dependent variable varies when a certain predictor variable changes, while holding all other variables nearly constant. In consequence, confounding due to association (correlation) with other variables is an omnipresent impediment to correct interpretation.

What we are looking at is not always what we are looking for.

region is sparsely settled, segments tend to be longer. Far-apart intersections and sparsely populated regions may have fewer driveways, more homogeneous speeds, more driver fatigue, be further from trauma centers, etc. Because of such associations, the relationship between segment length and accident frequency may be more complex than one of simple proportionality.

[19] A variable that has an important effect on the dependent variable but is not amongst the predictor variables.

3.6 Terrain

Having examined the relationship between $E\{\mu\}$, Segment Length, and AADT it remains to do an EDA for "Terrain." In the Colorado data "Terrain" comes in one of three categories: Flat, Rolling, and Mountainous; F, R, and M for short. The question is whether when AADT and Segment Length are held constant, the $E\{\mu\}$'s for one terrain differs systematically from the $E\{\mu\}$'s for another terrain. To answer, one has to create three tables similar to that in Fig. 3.12, one for each terrain. The pivot table makes the task easy.

Starting with the "Pivot Table Field List" which created Fig. 3.12 drag the "Terrain" field into the "Row Labels" window to be above the AADT 1994–1998 field. As shown on the left side of Fig. 3.15. This generates three pivot tables parts of which are shown on the right, one for each Terrain category (F, R and M).

As expected, comparing corresponding cells, the Rolling and Mountainous terrains have consistently more accidents per segment than the Flat terrain.[20] It follows that Terrain is a candidate predictor variable.

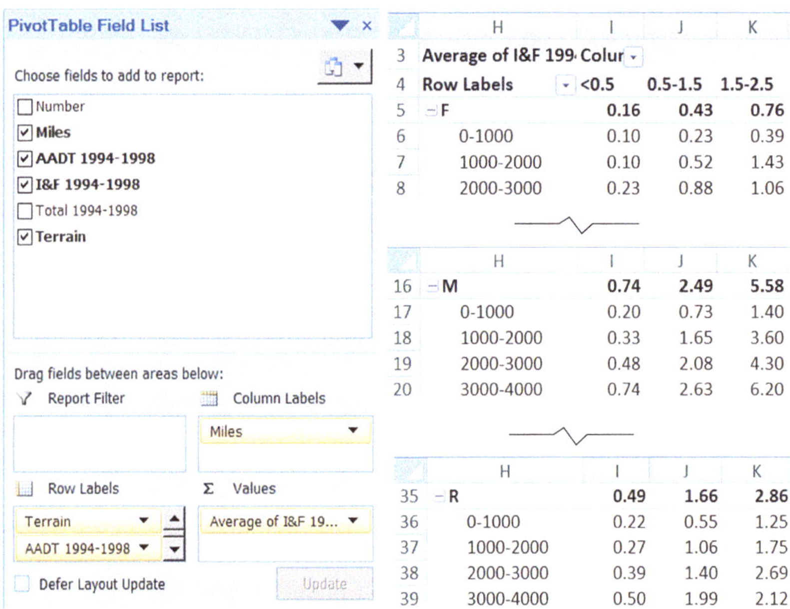

Fig. 3.15 The Pivot Table with Terrain categories

[20] The differences between F, R and M represent unaccounted for differences in grade, curvature, roadside, weather, road users, vehicles, etc.

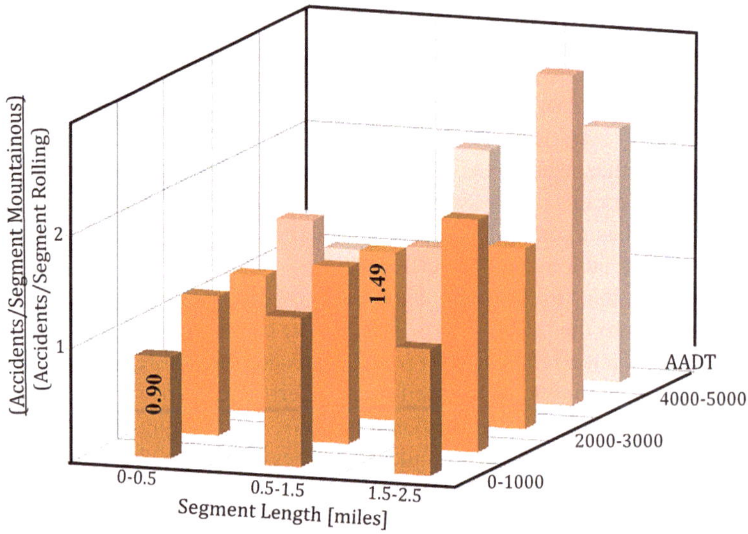

Fig. 3.16 How the ratio of Accidents/Segment depends on Segment Length and AADT

Less expected is the relationship in Fig. 3.16 which is the visual representation made of the results in Fig. 3.15. Consider, for example, the 0–0.5 miles and 0–1,000 vpd bins. Here $\hat{E}\{\mu_M\} = 0.20$, $\hat{E}\{\mu_R\} = 0.22$, and their ratio is $\hat{E}\{\mu_M\}/\hat{E}\{\mu_R\} = 0.90$. Compare this to the 0.5–1.5 miles and 2,000–3,000 vpd bins where the ratio is 2.08/1.40 = 1.49. The general impression is that the longer the segment and the more traffic it carries, the larger is the $\hat{E}\{\mu_M\}/\hat{E}\{\mu_R\}$ ratio. That means that the effect Terrain may depend Segment Length and AADT and this is how it may have to be represented in the SPF.

3.7 Summary

Forging a model out of data is a process that involves probing, computation, and the exercise of judgment. The latter requires, amongst other things, that the modeler develop a "feel" for the data. It is the role of Initial EDA to foster this feel. While this is a necessary early step in the modeling game, it is also a recurring one. Every addition of a variable to the SPF, and every choice of function will require an EDA-type activity.

The EDA story in this chapter is woven around data for Colorado two-lane rural roads. Data come with holes and errors. Ridding data of these is the first task of the initial EDA. What is fixed early saves tedious backtracking later. The spreadsheet is a friendly environment for data checking and cleaning.

The EDA helps answering two core questions. (a) Whether a trait is "safety related" and (b) what function can be used to represent it in the SFP. The Colorado data was used to answer these questions about Segment Length, AADT, and Terrain. To answer these questions the Pivot Table tool was introduced and used. This powerful spreadsheet tool greatly simplifies the conduct of an EDA.

The pivot tables were visualized in a series of graphs. In these both Segment Length and AADT were seen to be safety related. However, one could not confidently say what function links $E\{\mu\}$ and these two predictor variables. Terrain is also safety related. The impression is that its effect on $E\{\mu\}$ is a function of both AADT and Segment Length.

The numerical results of the EDA prompted three general observations.

1. The $E\{\mu\}$ of a population depends on the traits that define it. When a safety-related trait is added to those defining a population, the $E\{\mu\}$ changes.
2. For the $\hat{E}\{\mu\}$ of a population to be an unbiased estimate of the μ of a specific unit, the known traits of that unit must be the same as the traits which define the population.
3. When a trait is added to those defining a real population, the accuracy with which $E\{\mu\}$ is estimated diminishes.

It follows that the variables in the SPF must be the same as the data that are available for the practical task in which the SPF is used This has implications on how SPFs are to be developed and how reported.

References

Brillinger DR (2002) John Wilder Tukey. Not Am Math Soc 49(2):193–201

Chen Y, Persaud B (2014) Methodology to develop crash modification functions for road safety treatments with fully specified and hierarchical models. Accid Anal Prev 70:131–139

Chiou Y-C, Fu C (2013) Modeling crash frequency and severity using multinomial-generalized Poisson model with error components. Accid Anal Prev 50:73–82

Gross F, Craig L, Persaud B, Srinivasan R (2013) Safety effectiveness of converting signalized intersections to roundabouts. Accid Anal Prev 50:234–241

NIST/SEMATECH e-handbook of statistical methods. http://www.itl.nist.gov/div898/handbook/

Harwood DW, Torbic DJ, Richard KR, Meyer MM (2010) SafetyAnalyst software tools for safety management of specific highway sites. FHWA-HRT-10-063, Federal Highway Administration, Office of Safety Research and Development

Curve-Fitting

4

Abstract
Curve-fitting[1] is based on the belief that hidden under the data cloud there is an orderly relationship between the $E\{\mu\}$ and the population-defining traits. The key feature of curve-fitting is that use is made of data from neighboring populations to fashion estimates for specific populations. This is intended to alleviate the "sparse data problem." But curve-fitting is not an unmixed blessing. As it irons out unwanted randomness, it distorts some of what is real. When curve-fitting is "nonparametric," one only has to specify a rule by which an estimate is to be computed from neighboring data. For "parametric" curve-fitting, the modeler has to assume that the underlying orderly relationship can be represented by some equation, and then to estimate its parameters. To illustrate the nature of nonparametric curve-fitting, the "Nadaraya-Watson Kernel Regression" will be applied to the Colorado data.

4.1 Why Do We Need to Curve-Fit?

The purpose of an SPF is to provide estimates of $E\{\mu\}$ and $\sigma\{\mu\}$ for a multitude of populations as a function of population traits and their levels.[2] The Initial EDA gave hints about whether and how accidents on rural two-lane roads in Colorado depend

[1] The Wikipedia describes curve-fitting as the process of constructing a curve or mathematical function that has the best fit to a series of data points. The term "curve-fitting" tends to be used by engineers and scientists whereas regression analysis is the preferred term for statisticians and social scientists.

[2] Some believe that the purpose of SPF development goes beyond the provision good estimates of $E\{\mu\}$ and of $\sigma\{\mu\}$, that the objective is to determine by how much will $E\{\mu\}$ change if the level of some trait is purposefully changed. Whether their goal is attainable is subject to debate (See, e.g., Hauer (2010), Elvik (2011)). More on this is in Sects. 6.7 and 10.2. Be it as it may, the SPF development described in this book is guided by the need for good estimates of $E\{\mu\}$ and $\sigma\{\mu\}$ as

E. Hauer, *The Art of Regression Modeling in Road Safety*,
DOI 10.1007/978-3-319-12529-9_4

on Segment Length, AADT, and Terrain. However, most bins in Fig. 3.10 contained too few segments for meaningful estimation. Thus, even with a large data set and few population-defining traits the information in most bins is too slim to be of use. This is the "sparse data problem."[3] The saving grace is the belief that some traits are "safety-related"[4] and exhibit an "orderly relationship"[5] with what is being estimated. This orderliness can be exploited for estimation because it means, in essence, that what is observed in one bin contains useful information about what to expect in a nearby bin.[6]

To illustrate, the pivot table was used on the condensed Colorado data[7] to produce columns 1–4 in Table 4.1. Column 5 is a five-point running average of the entries in column 4. Thus, e.g., the ordinate at, say, $7{,}000 < \text{AADT} \leq 7{,}500$ is $3.30 = (2.37 + 3.37 + 6.15 + 2.64 + 1.98)/5$.

That means that instead of using the information about the 41 segments in the $7{,}000 < \text{AADT} \leq 7{,}500$ bin and claiming that $\hat{E}\{\mu\} = 6.15$, we prefer to rely on the smoother curve made of the running averages (see Fig. 4.1) and to believe that $\hat{E}\{\mu\} = 3.30$ is the better estimate.

By making the estimate into a function of neighboring data, the "sparse data" curse can be partly lifted. However, as with all curses, the remedy requires faith; here it is faith in the existence of an underlying orderly relationship. It is through this act of faith that "curve-fitting" becomes the kingpin of modeling.

Curve-fitting is of two kinds: "nonparametric" and "parametric." When curve-fitting is nonparametric, the notion of an orderly relationship is limited to the belief

---

these are used in practical applications discussed in Sect. 1.3, illustrated in Sect. 1.4, and described in the Highway Safety Manual (AASHTO 2010).

[3] Many phenomena in diverse domains (numerical analysis, sampling, combinatorics, machine learning, data mining) are referred to by this name. The common theme of these is that when "dimensionality" increases, the "volume of the space" increases so fast that the available data becomes "sparse." Dimensionality here refers to the number of traits. Thus, with two traits, say Segment Length and AADT, we have two "dimensions" and the "volume of space" is a table or a plane graph. The addition of Terrain makes for three dimensions and the volume of space is now a rectangular prism or three dimensional graph. Sparsity is a problem because to obtain statistically sound results, the amount of data needed grows exponentially with the dimensionality. The phrase "curse of dimensionality" was coined by Richard E. Bellman when considering problems in dynamic optimization.

[4] A trait was said to be safety related if changing it changes the μ of units.

[5] A relationship was said to be orderly if it exhibits a pattern of data points that makes the fitting some curve to it a sensible choice. This is a troubling definition because it amounts to saying that a relationship is orderly if it seems to be so. An orderly relationship is but a promise of safety relatedness, not a guarantee of it.

[6] The notion of "useful information" contains a seed of an idea which may help doing away with the circularity in the definition of "orderly relationship." The observed values in bin X are useful for estimating some unknown parameter in bin Y if using them improves the quality of the parameter estimate. A relationship is orderly if the preceding sentence is true for most bins. More about this is in Sect. 4.2.

[7] File "4a or 4b Colorado condensed.xls or xlsx" in the "Data" folder from http://extras.springer.com/ and the ISBN of this book.

Table 4.1 Five-point running averages

1	2	3	4	5
AADT range	No. of segments	No. of I&F accidents	Average I&F accidents/segment	SPF ordinate: five-point running average
0–500	808	473	0.59	
500–1,000	1,176	929	0.79	
1,000–1,500	615	902	1.47	1.15
1,500–2,000	476	650	1.37	1.48
...	...	...	...	...
6,000–6,500	62	147	2.37	3.34
6,500–7,000	57	192	3.37	3.35
7,000–7,500	41	252	6.15	3.30
7,500–8,000	61	161	2.64	3.59
8,000–8,500	41	81	1.98	3.77
8,500–9,000	29	111	3.83	3.11
9,000–9,500	30	128	4.27	3.33
...	...	...	...	...
18,000–18,500	1	14	14.00	
19,500–20,000	2	15	7.50	

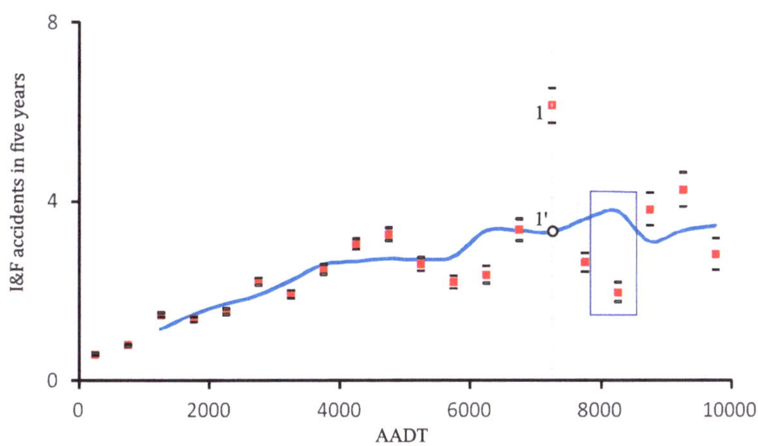

Fig. 4.1 Problems with nonparametric fits

that the estimate at a certain level of a trait can benefit from data in the neighborhood of that level. One only needs to specify the corresponding computational rule. The running average is one such rule. The products of nonparametric curve-fitting are tables and graphs, not equations.

Nonparametric curve-fitting is a rule for computing an estimate using both local and neighboring data.

When curve-fitting is "parametric" one must assume more. Now the assumption is that hidden under the fog of data there is some smooth curve or multidimensional surface and that it can be adequately represented by a simple mathematical expression. That expression is called the "model equation." The "model equation" consists of traits (variables), and of parameters.[8] The modeler then has to decide which variables to use, to specify the function whereby these variables are to be combined, and then to estimate the parameters of the model equation.[9] Once the model equation is specified and its parameters estimated we have an SPF.

Parametric curve-fitting is based on a "model equation" made up of "variables" and "parameters."

This chapter deals only with nonparametric curve-fitting. Curve-fitting of the parametric kind is the subject of the remaining chapters.

4.2 There is No Free Lunch

The essence of curve-fitting is that it brings data from neighboring populations to bear on the estimates of $E\{\mu\}$ and $\sigma\{\mu\}$ for a specific population. The price of this stratagem is the loss of what is unique about a specific population and the importing into it of "foreign matter" from the neighborhood. These issues are illustrated in Fig. 4.1.

The figure is based on the Colorado data used earlier. As before, the ordinate of a square is the average number injury and fatal (I&F) accidents in 5 years per segment in an AADT bin[10]; the horizontal bars which surround the squares are one standard

[8] The general notation for the model equation of an SPF is $E\{\mu\} = f(X_1, X_2, \ldots, \beta_0, \beta_1, \beta_2, \ldots)$ where $f()$ denotes some function of the predictor variables $X_1, X_2, \ldots$ and of the parameters $\beta_0, \beta_1, \beta_2, \ldots$; the $E\{\mu\}$ is the "dependent variable." In statistics it is common to call it the "response variable." This terminology is tendentious because it invites the interpretation that one can use the model equation for determining the "response" of $E\{\mu\}$ to a change in an X. As argued in Sect. 6.7, such a use of SPF is fraught with problems. For this reason, I prefer to call $E\{\mu\}$ the "dependent variable."

[9] Parameters are constants that complete the determination of the mathematical expression in which the variables are combined. Thus, for example, in $f(x) = \beta x$, the parameter "β" determines the slope of the line described by $f(x)$. Learned papers treat the parameter estimation topic with much relish. It is as if the trustworthiness of the model equation was determined mainly by the accuracy of its parameters. The unadorned fact is that there is little data-based theory behind the choice of traits (variables) that are used in the model equation and little attention given to the way in which they combine into the mathematical expression for which the parameters need to be estimated. Why the focus on parameters and their estimation and the lack of interest in the expression of which they are a part is perhaps explainable as a part of the history of statistics; it is nevertheless difficult to justify and accept.

[10] The AADT bins here are 500 vehicles per day wide.

error away. The fitted smooth curve is based on a five-point moving average.[11] Thus, for example, the ordinate of the curve at point 1′ is the arithmetic average of the ordinate at point 1 and of the four squares nearest to it. There are, of course, more elaborate methods of nonparametric curve-fitting but my argument is best made by what is simple and familiar.

The ordinates of points 1 and 1′ are quite different. Considering the ± standard error shown by the horizontal bars which surround the square at 1, the difference between the two ordinates cannot be attributed statistical inaccuracy; there is something about the road segments in the $7{,}000 < \text{AADT} < 7{,}500$ bin that is different from the segments in the four surrounding bins. If so, it may be quite unreasonable to estimate the $E\{\mu\}$ of a segment with AADT = 7,250 by the ordinate of the curve at 1′. And yet, that is what is always done when a curve is fitted to data. The moral is that as it smoothes out randomness, the fitting of a curve also irons out some real differences; no curve-fitting method can avoid doing so.

Curve-fitting gets rid of some unwanted randomness but erases some of what is real and should be kept.

Consider now the kink of the fitted curve inside the rectangle. The kink is due entirely to the unusually large ordinate of point 1 which influences the running averages inside the rectangle. There is nothing in the data within the rectangle to suggest or to justify such a peculiar upswing of the curve. This distortion is also a cost of curve-fitting. Inasmuch as the estimate for a specific population is made into a function of neighboring data, it is inevitable for "foreign matter" to exert influence on every local estimate.

Yet another problem is visible in Fig. 4.1. Judging by the horizontal bars, up to AADT of about 5,000, the ordinates of the squares are fairly accurately estimated. Is the smooth curve fitted to this domain a better estimate of $E\{\mu\}$ than the squares? There is nothing intrinsically superior about a smooth curve nor is there reason to believe that the underlying curve is smooth. Inasmuch as geometric design standards and road maintenance policies are defined by a few discrete AADT ranges, one may expect discontinuities at the breakpoints. It is therefore unclear whether the reliance on a smooth curve improves estimation or degrades it. It may not be an improvement to impose a smooth curve onto the data, especially not in regions where bin-based estimates are accurate.

In sum, the fitting of some nonparametric smooth curve to data is associated with obvious problems: it tends to erase some of what is real, it creates estimates that are

[11] Given a series of numbers and a "window" (subset) size, the moving average can be obtained by computing the average of the values in the window. The window is then shifted forward, creating a new subset of numbers, which is averaged. This process is repeated over the entire data series. In this illustration the average is of five ordinates. Between AADT = 0 and 10,000 with twenty data bins of 500 vpd, there are $20-4=16$ running average estimates. The Excel graphing option of "Scatter with smooth line" makes a smooth curve out of the 16 estimates.

not in line with what was observed, and it imposes a smooth relationship on a reality that is possibly discontinuous. Why then is smoothing attractive? The attractions are two. First, as will be shown shortly, it succeeds in identifying order even when it is not visually discernible. Second, it appeals to our predisposition to think that "Natura non facit saltus[12]"; to believe that the relationship in which we put our faith is in fact smooth.

When curve-fitting is "parametric," all the weaknesses of nonparametric curve-fitting are retained and another is added. Now the modeler must choose a "model equation" to represent the relationship assumed to hide within the data. While that relationship can take on any form, the modeler has to commit to a specific equation.[13] Doing some severely restricts the range of possible shapes. Commonly used model equations can render only simple shapes. Most cannot even have a peak, a valley, or an inflection point, not to speak of undulations such as those in Fig. 4.1. Thus if there is any richness or nuance to the relationship, the chosen equation will distort reality.[14]

Model equations are straightjackets for data.

Thus, if nonparametric curve-fitting could do the job, it should be the tool of choice. Can it do the job? To gain insight I will examine some nonparametric curve-fits to the Colorado data.[15]

4.3 Kernel Regression

The moving average method of curve-fitting which generated the curve in Fig. 4.1 has an obvious fault; it does not allow for a nearby ordinate to exert a larger influence over the estimand than one that is further away. The "Nadaraya-Watson Kernel Regression" to be used here is also a moving average method of smoothing except that close neighbors are given more weight than distant ones.[16] Here the weights will be derived from a Normal "bell curve" the middle of which is placed

---

[12] Natura non facit saltus (Latin for "nature does not make jumps") has been a principle of natural philosophy since at least Aristotle's time.

[13] A model equation is what in mathematics is called a "family of curves"; a set of curves whose equations are the same but which have different values assigned to one or more of their parameters.

[14] To illustrate Lord and Bonneson (2007), chose to represent the effect of lane width by the function $\exp(\beta \times \text{Lane Width})$ and estimated β to be -0.188. This function can have no minimum and this is why they conclude that as lane width increases from 9′ to 13′ accident frequency continues to decrease. Early empirical evidence suggests that for rural two-lane roads there is a minimum somewhere between 11′ and 12′, that widening beyond 12′ may increase accident frequency. Such a conclusion is unreachable once the $\exp(\beta \times \text{Lane Width})$ function is adopted.

[15] To access and download, the data go to http://extras.springer.com/ and enter the ISBN of this book. Look in the "data" folder for files 4 (a or b) Colorado condensed. "(xls or xlsx)"

[16] See Nadaraya (1964), Watson (1964) or, e.g., Simonoff (1996), Li and Racine (2007).

above the point at which the weighted average is to be calculated.[17] In this context the standard deviation of the bell curve is called the "bandwidth."

The 5,323 data points[18] are shown in Fig. 4.2. In this representation no underlying pattern is discernible and the question is whether the nonparametric wizardry can find order hidden under this cloud.

The Nadaraya-Watson algorithm has been coded onto a spreadsheet[19] so that the user can fit a nonparametric curve to any data by a click of a spreadsheet command button. Figure 4.3 shows the fitted SPFs for bandwidth = 200 (solid) and 1,000 (dashed) against the background of the data points.

The merit of nonparametric curve-fitting is now clear; it lifted from the data cloud an otherwise un-discernible relationship. The question is which of the two fits shown or, equivalently, which of the two bandwidths is to be preferred. The issues arising require airing.

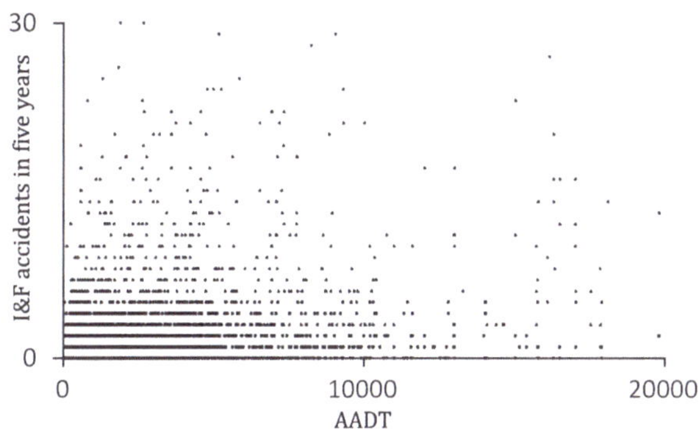

Fig. 4.2 The data cloud

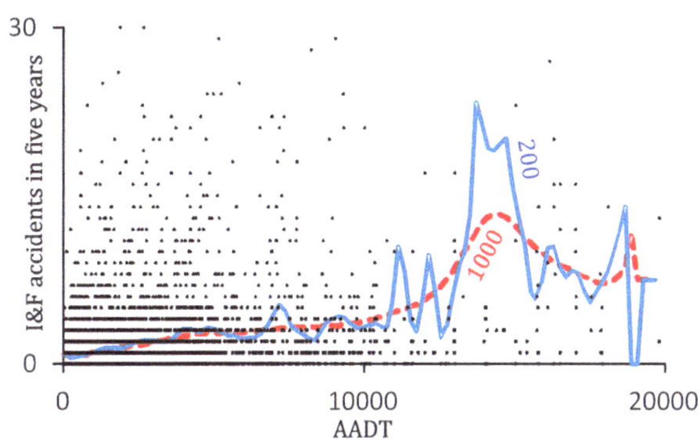

Fig. 4.3 Nadaraya-Watson fits with two bandwidths

[17] For detail, see Appendix H.

[18] Points with more than 60 accidents were omitted from the figure.

[19] For more detail about the N-W algorithm is in Appendix H. To download the spreadsheet, go to http://extras.springer.com/ and enter the ISBN of this book. Look in the "Spreadsheets" folder for "Chap. 4. Nadaraya-Watson non-parametric.xlsm".

4.3.1 Bandwidth and Goodness of Fit

The number of values used in computing a running average (it was five in Table 4.1) and the bandwidth used in the Nadaraya-Watson algorithm are affine concepts. The smaller the number of values averaged or the narrower the bandwidth, the less weight goes to more distant neighbors, the more undulating is the curve, and the closer it approaches the data points. Conversely, the wider the bandwidth the more damped are the oscillations of the curve and the further it is from the data points. It follows that the curve generated using the lesser bandwidth in Fig. 4.3 has a better "goodness of fit" measure[20] than the curve generated using the wider bandwidth.

The notion that a broadly oscillating curve always fits the data better than a mildly undulating one must raise eyebrows for it is destructive of the very belief that the relationship is orderly. There surely is a bandwidth beyond which a further narrowing of it will degrade the correspondence between the fitted curve and a plausible reality. If so, the proximity between fitted values and data points – the goodness of fit measure – fails as a measure of fit quality; an alternative yardstick must be sought.

To stay on solid ground, it is best to go back to the purpose of curve-fitting – that of providing good estimates. To explain, consider again the data in Table 4.1. If the average of the two closest neighbors was the chosen estimator then, for the $7{,}000 < \text{AADT} < 7{,}500$ bin, the estimate would be $(3.37+2.64)/2=3.00$. The difference between this estimate and the sample mean computed from the bin data alone (6.15) would be 3.15. If the estimator was the average of the four closest neighbors, then the estimate would be $(2.37+3.37+2.64+1.98)/4=2.59$ and the difference would be $6.15-2.59=3.56$. Proceeding similarly for six and eight closest neighbors, the ranking in terms of |difference| of the four estimators is in Table 4.2. For this AADT bin, the "two-closest-neighbors" happens to be best.

A similar ranking can be obtained for all the bins in Table 4.1. For this data, the eight-closest-neighbors estimator has the lowest average rank and is therefore considered best. This rationale comes from the answer to the question of which estimator, had it been used, would have come closest to what has been observed (but not used in the estimator). A similar rationale can be used for determining the optimal bandwidth.

Table 4.2 Rank of four estimates for the $7{,}000 < \text{AADT} < 7{,}500$ bin

Estimator	SPF ordinate	Absolute difference with 6.15	Rank
Two closest neighbors	3.00	3.15	1
Four closest neighbors	2.59	3.56	4
Six closest neighbors	2.73	3.42	3
Eight closest neighbors	2.91	3.24	2

[20] Many goodness of fit measures are in use. One such measure is the sum of squared differences between a fitted curve and the data points.

To illustrate, 1,000 of the 5,323 Colorado segments were selected at random. Using varying AADT bandwidths (100, 150, 200, 300, 500, and 1,000 vpd) curves were fitted by the N-W algorithm to data for the remaining 4,323 segments. The proximity of the estimates obtained by the different bandwidths to the observed number of accidents determined the rank of each estimate for all 1,000 segments. From these the average rank for each bandwidth was computed.[21] The result is in Fig. 4.4.[22]

An indifferent estimator (or bandwidth), one which is equally likely to assume any rank, would have an average value of (1 + number of estimators ranked)/2. With different bandwidths tested here the indifference line is at 3.5. The best bandwidth in this case is about 200 vpd.

At first glance this result may be thought surprising because it favors the more turbulent curve in Fig. 4.3. The surprise is rooted in our predisposition to believe that $E\{\mu\}$ is some smooth function of AADT and that there is no reason why the relationship should be a wavy one. On second thought, recall the earlier dictum that in a

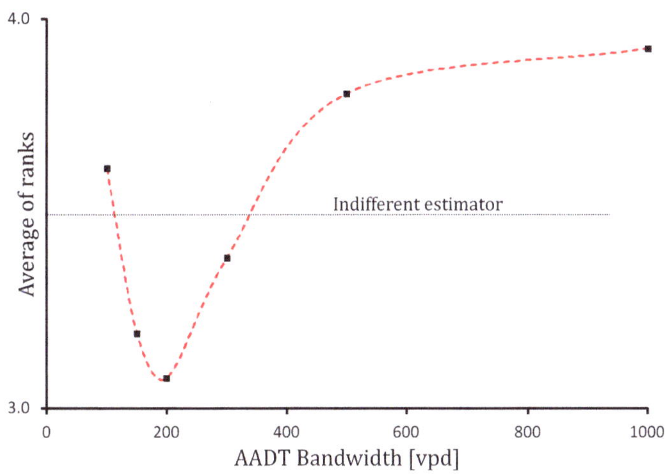

Fig. 4.4 Average rank as a function of bandwidth.

[21] This approach to bandwidth selection is referred to in the nonparametric regression literature by the name "leave-one-out cross validation" (LOOCV). Inasmuch as the central idea is that of determining which bandwidth would predict best for those observed values that were not used in generating the fitted value, in the present context a better term might be "prediction optimization." To download the spreadsheet, go to http://extras.springer.com/ and enter the ISBN of this book. Look in the "Spreadsheets" folder for "Chap. 4. Nadaraya-Watson non-parametric.xlsm".

[22] This comparison covers the range of $3{,}000 < \text{AADT} < 8{,}000$. For $\text{AADT} < 1{,}000$ estimates could not be obtained with a bandwidth of 1,000. For estimates with $\text{AADT} > 8{,}000$, the bins do not contain enough segments for the comparisons to be reliable.

two-dimensional graph, "what we are looking at is not always what we are looking for." In Fig. 4.3 we look at the influence of AADT but must remember that Segment Length, Terrain, and of a host of other variables remain unaccounted for. Therefore, if a cluster of data points is above the fitted curve it may be, partly, because these segments that are longer than average or are mainly in mountainous terrain. This may explain why the best-predicting curve is wavy; the undulations in it correctly reflect the influence of unaccounted-for variables. Several observations follow.

First, when the best fitting curve is oscillating while a smooth curve should be expected, it is a sign that the influence of some safety-related variables has not yet been accounted for. Second, if a-priory considerations point to a smooth relationship but the best-predicting curve is wavy, the latter should be used for prediction. Third, using a smooth curve for prediction when not all safety-related variables are accounted for is not in the interest of good estimation or prediction. Fourth, when a smooth function is expected and the best-predicting curve is smooth, it is a sign that a satisfactory result has been attained. To reduce unwanted oscillations and to attain a satisfactory result safety-related variables should be added. How this can be done is examined next.

4.3.2 Adding a Variable

In the two-dimensional case the weights for the Nadaraya-Watson kernel regression were derived from a univariate Normal "bell curve" the middle of which was placed above the point at which the weighted average is to be calculated. The standard deviation of the bell curve was the "bandwidth." To add a variable, one has to specify a bivariate bell curve[23] which has two standard deviations,[24] i.e., two bandwidths.

With one predictor variable, AADT, the Nadaraya-Watson procedure succeeded in bringing out a relationship that would otherwise not be discernible. The question is whether similar success can be attained when more variables are used. In Fig. 4.5 Segment Length (L) was added to the AADT variable. For AADT the bandwidth was 500 vpd, for L (Segment Length) a bandwidth of 0.5 miles was used.

Even with relatively wide bandwidths, the region where the fit is stable is limited to where data is plentiful, the identified as such in Fig. 3.10. When AADT is larger than about 8,000 and for segments longer than 2–3 miles, the undulations of the fitted curves begin to wash over the underlying regularity. Thus, while curve-fitting was deployed to alleviate the sparse data problem, with only two variables used, the

[23] For detail, see Appendix H.

[24] The correlation coefficient of the bivariate bell curve is set to 0.

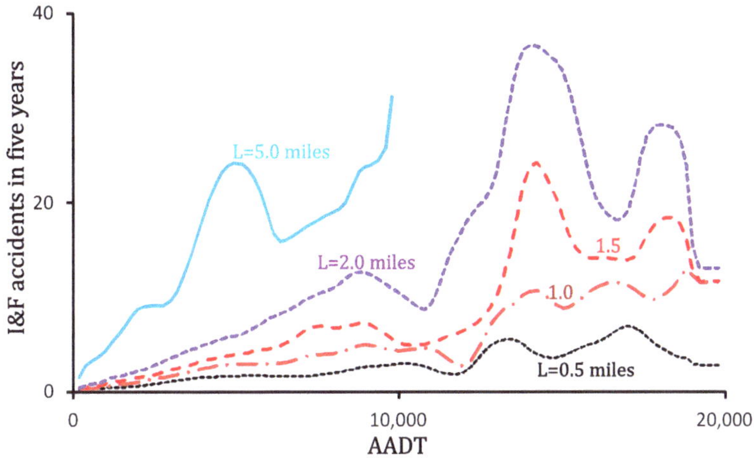

Fig. 4.5 Nadaraya-Watson regression with two variables. $\text{Bandwidth}_{\text{AADT}} = 500$ vpd, $\text{Bandwidth}_{\text{L}} = 0.5$ miles

dimensionality curse is seen to raise its head again.[25] It seems that nonparametric curve-fitting can only help with SPFs with one or two traits. For SPFs with more than two traits nonparametric fits are useful mostly as a tool of the EDA.

The main merit of nonparametric regressions is that the modeler does not have to declare what mathematical expression hides behind the data. It seems now that in a world in which the safety of units depends on more than one or two traits the making of such heroic assumptions cannot be avoided. By trial and error, intuition and experience the modeler must gradually give shape to a model equation that, with luck, might adequately replicate the main features of the underlying regularity. In doing so one has to say which traits (variables) should feature in the model equation, in what manner they should combine into an equation, and what should be the values of the unknown parameters in that equation. The traits chosen to be included in the model equation and the way in which they combine, give the model equation its range of possible shapes; they are the skeleton and lattice of the assumed regularity. The parameter values, in turn, determine how this skeleton and lattice are stretched and bent to bring them into conformity with the data. This stretching and bending invariably involves some kind of minimization or

[25] Viewed in an algebraic light, adding a new trait with "N" levels divides the average number of observations per each cell by "N." With such a geometric progression the information in the cells of even the richest data set is quickly decimated into uselessness. Therefore, even though multidimensional non-parametric curve-fitting is conceptually feasible, its practicality is destroyed by the rapid diminution of data caused by the addition of traits.

maximization[26]. Thus, for example, one may aim to find parameters that minimize the sum of squared differences or of the absolute differences between the data points and the assumed shape. Alternatively, one may want to identify that set of parameter values which maximizes the probability to observe the available data. The function to be minimized or maximized is called the "objective function." The question of what objective function to use will be discussed later in Sect. 8. To prepare for estimating the parameters of model equations one has to know how to find extrema of functions using a spreadsheet. This is the topic of the next chapter.

Parameter estimation always requires the minimization or maximization of some "objective function."

4.4 Summary

Curve-fitting is the set of activities that transforms data into an SPF. It rests on the belief in the existence of an "orderly relationship" between what is being estimated and the level of traits. This belief is what allows one to bring data from neighboring populations to bear on the estimate for a specific population. One hopes that curve-fitting alleviates the "sparse data problem."

The assumed-to-exist orderly relationship and the use of neighboring data are responsible for the shortcomings of curve-fitting; they diminish what is unique about the population of interest. Moreover, a smooth-fitted curve may not make for better estimates then bin estimates when bins are narrow and data plentiful.

There are two kinds of curve-fitting. When curve-fitting is "nonparametric," one only has to specify a rule by which an estimate is to be made up from the data. The resulting SPF is a table or a graph. When curve-fitting is "parametric," the modeler has to choose a mathematical expression, a model equation, and then estimate its parameters. Now the SPF is an equation. The use of a model equation exacerbates the blemishes of nonparametric curve-fitting. There is no reason to think that the complexities of accident generation that are hidden within the data can be represented by some simple assumed mathematical expression. In sum, while perhaps unavoidable, curve-fitting is not an unmixed blessing.

To illustrate the nonparametric curve-fitting, the "Nadaraya-Watson Kernel Regression" was applied to the Colorado data. A code was written so that the user can fit a nonparametric curve to any data by a click of a spreadsheet command

[26] Minimization is the mirror image of maximization. To convert one into the other all that has to be done is to put a minus sign in front of an expression. By doing so the task of finding the deepest valley turns into the task of finding the highest peak. Therefore "optimization" will be used to mean either the task of minimization or that of maximization. The common term for minima and maxima of functions is "extrema".

button. The result can be quite impressive when it brings out the image of an underlying regularity which is not apparent by just looking at a graph of the data.

What curve is fitted and what values it predicts depends on what bandwidth is chosen. To aid this choice, a method to find the bandwidth that yields the best-predicting fitted curve was presented. Contrary to what may be expected the best curve was wavy. The presence of undulations when a smooth curve is expected is a sign of unaccounted for safety-related variables.

A nonparametric fit works well with one trait but its attraction diminishes when two or more traits have to be used. Thus, example, as soon as Segment Length was added as a trait to AADT the undulations of the fitted curves began to obscure the underlying regularity. When several traits are thought to affect the SPF estimates, nonparametric fits are useful mostly as EDA études.

References

AASHTO (The American Association of State Highway and Transportation Officials) (2010) Highway Safety Manual, 1st edition

Elvik R (2011) Assessing causality in multivariate accident models. Accident Anal Prev 43:253–264

Hauer E (2010) Cause, effect, and regression in road safety: a case study. Accident Anal Prev 42:1128–1135

Li Q, Racine JS (2007) Nonparametric econometrics: theory and practice. Princeton University, Princeton

Lord D, Bonneson JA (2007) Development of accident modification factors for rural frontage road segments in Texas. Transport Res Rec 2023:20–27

Nadaraya EA (1964) On estimating regression. Theor Probab Appl 9(1):141–142

Simonoff JS (1996) Smoothing methods in statistics. Springer, New York

Watson GS (1964) Smooth regression analysis. Sankhya Ser A 26:359–372

Preparing for Parametric Curve-Fitting: The "Solver" 5

Abstract
Parametric modeling requires parameter estimation which, in turn, requires optimization. For study, for instruction, and for practical road safety modeling one can use an optimization tool available in Excel: the Solver.

The use of the Solver is explained and its application to curve-fitting illustrated. While the Solver is a robust tool two problems may cause it to fail. First, while Solver is good at converging on local optima, finding the global optimum is a challenge. The "correspondence graph" can help alleviate this problem. Second, the algorithm may fail for computational reasons when the parameters to be estimated differ by several orders of magnitude. This problem may be obviated by appropriately scaling the model variables.

5.1 Optimization in Modeling

Nonparametric curve-fitting does not produce useful SPFs when two or more variables have to be used. Because the safety of units depends on several safety-related variables, the need to specify a model equation and to estimate its parameters cannot be further avoided. This is why from here on, the subject matter is that of parametric curve-fitting.

One seeks to find parameter values that make for a best fit or produce best estimates. When "best" is said optimization is implied; it may be the minimization of the sum of squared differences, the maximization of likelihood, the minimization of absolute residuals, etc. For all these an optimization tool is needed.

When parameters are many, when they combine with variables in nontrivial functions, and when datasets are large, it is not simple to find the optimizing set of parameter values. Optimization routines are a part of statistical packages; these can be purchased and, with some effort, mastered. However, for instruction, study, and for most practical tasks, the use of the Solver optimization tool in Excel is

E. Hauer, *The Art of Regression Modeling in Road Safety*,
DOI 10.1007/978-3-319-12529-9_5

sufficient, perhaps even preferable.[1] In this chapter the Solver tool will be introduced and its uses illustrated. After that the modeling adventure will begin.

5.2 Using the Solver to Find Minima and Maxima

To illustrate the use, capabilities, and the limitations of the Solver[2] let us find the peaks and the valleys of the objective function[3] in Fig. 5.1.[4]

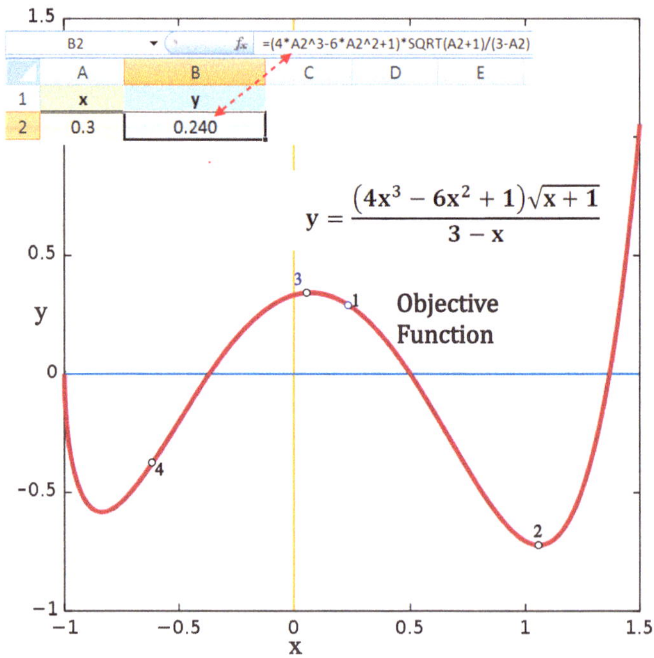

Fig. 5.1 Finding the extrema of an objective function

[1] The Solver comes with Excel but has to be installed and referenced. Installation instructions may be found at http://office.microsoft.com/en-us/excel-help/introduct. Before its first use the "Excel Solver" has to be "referenced." To do so, go to "Developer," on the "Code" tab go to "Visual Basic," and in the window that opens click on "Tools." On the menu select "References," check the "Solver" box, and then click OK.

[2] For comprehensive guidance on Solver see Harmon (2011). For a description of its development and use see Fylstra et al. (1998).

[3] The term "objective function" is associated with some real setting in which an increase in the function is thought either desirable or not desirable. This is why there is interest in finding that combination of variables which make the objective function either largest or smallest.

[4] To download this spreadsheet go to http://extras.springer.com and enter the ISBN of this book. Look in the "Spreadsheets" folder for "Chap. 5. How to use the Solver.xlsx."

To begin enter some value for x in one cell of a spreadsheet (here $x = 0.3$ in cell A2) and the formula for y as a function of that value in another cell (here the formula is in cell B2). The 0.240 which the spreadsheet shows in B2 is the ordinate of point 1 in Fig. 5.1. With this simple machinery one could find the peaks and valleys of the function by intelligently entering value after value in A2. However, if y was a function of several variables the task would be tedious. The Solver does the work efficiently.

On the top left part of the Excel (2007) spreadsheet in Fig. 5.2 click on the "Data" tab and then, on the top right, click on "Solver." The "Solver Parameters" window opens.

The aim is to find the extrema of "y" and thus the "Target Cell" of the optimization is cell B2.[5] The "radio buttons" are used to tell the Solver whether to find a minimum or a maximum. Thus, if we want to find a minimum, the "Min" button has to be clicked. Finally, the value in the Target Cell is minimized by changing the value of x and therefore A2 goes into the "By Changing Cells" window. With these three settings click the "Solve" button in the upper right corner.

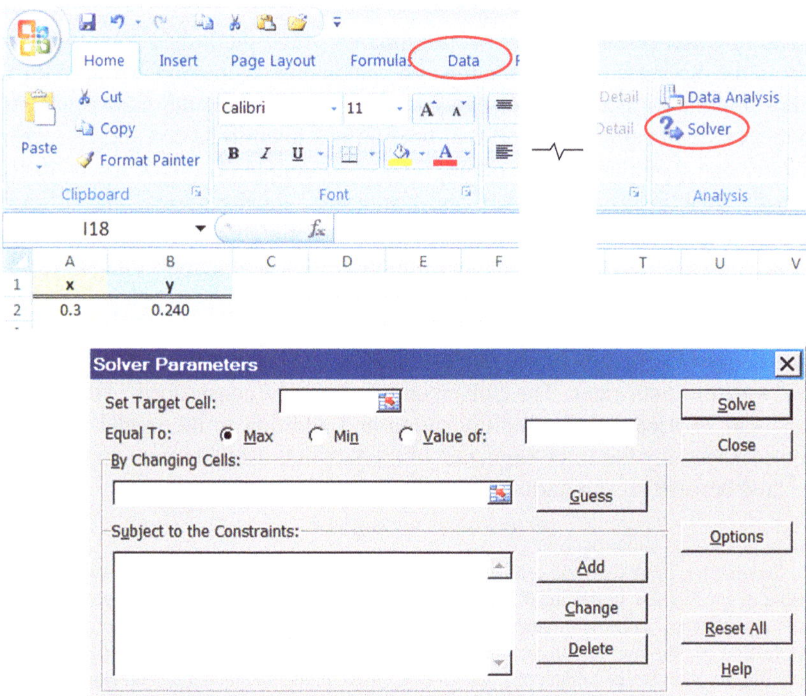

Fig. 5.2 Opening the "Solver Parameters" window

[5] The "Target Cell" in Solver is equivalent to "Objective Function" in optimization.

The Solver finds a minimum at $x = 1.058$ in cell A2 and $y = -0.723$ in cell B2. These coordinates correspond to point 2 of Fig. 5.1. Were the radio button set to "Max" (and starting again from $x = 0.3$) the Solver would find the maximum at point 3 ($x = 0.070$, $y = 0.343$).

The main features of the Solver are now apparent. The search for an extremum begins from an "initial guess"; here it was the 0.3 in A2. At this point Solver computes the direction of the largest slope. If instructed to search for a minimum it takes a step downhill in that direction; if asked to find the maximum it takes an uphill step. Solver repeats the same process (each time adjusting the step size) till it reaches a point from which further descent is not possible. It follows that Solver can only find the minimum that is downhill from the initial guess. Similarly, if instructed to search for a maximum, it can only find that peak that is reachable by climbing uphill from the initial guess.

The Solver finds only one extremum; usually that which is nearest the initial guess.

The function in Fig. 5.1 has two minima of which the Solver found only one. This is the Solver's main limitation. For parameter estimation we need to find global minima or maxima and one cannot know whether the extremum which Solver finds is local or global. If the function is known to be convex, i.e., when it has only one minimum or maximum, the task is simple. The task can be difficult when the function is non-convex and has several local extrema. The difference between a convex and a non-convex function is depicted in Fig. 5.3.[6]

Because the search for the global optimum of a non-convex function can be troublesome some statistical packages may shun model equations and likelihood structures that lead to non-convexity. To avoid this awkward problem it is tempting to use convex but inappropriate model equations. This amounts to throwing out the baby with the bathwater. The Solver is not similarly constrained. All functional forms can be tried; mixing multiplication and addition in the model equation is allowed. One only has to be cognizant of the fact that the extremum may be a local one[7] and beware of singularities.[8]

---

[6] If a function has more than one peak or valley they are its "local extrema." The lowest of the local minima or the highest in the local maxima is the "global optimum." When the parameters of a model equation are to be estimated we are interested in the global optimum. The task of global optimization of non-convex function remains difficult. The search for efficient algorithms to find global optima is an active branch of research in applied mathematics. An e-sourcebook of what is known is by Weise (2008). The standard Microsoft Excel Solver and Premium Solver Pro do not offer built-in facilities for solving global optimization problems.

[7] To increase the Solver's chance of converging on a globally optimal solution the following two-step strategy can be tried: (1) Begin the curve-fitting by minimizing the sum of squared differences which is known to be convex. (2) Use the solution from step 1 as a starting point for any other objective function.

[8] In mathematics a singularity is a point at which a given mathematical object is not defined or not well-behaved.

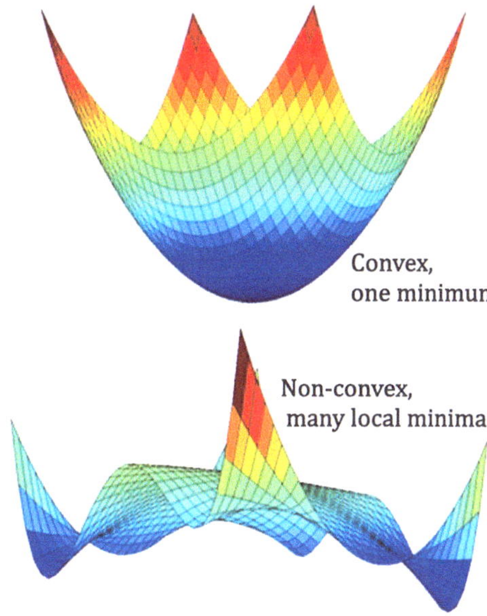

Fig. 5.3 Convex and non-convex functions

Having identified a local minimum at point 2 and a local maximum at point 3 in Fig. 5.1 there may be another local minimum on the other side of this peak. To find it one can choose an initial guess a little to the left of the peak at, say, $x = 0.05$, place this value in A2, set the radio point to "Min," and click "Solve." The result is in Fig. 5.4.[9]

What went wrong? There is good reason why the graph in Fig. 5.1 does not extend to values to the left of -1; that is where the radical $\sqrt{(x+1)}$ turns imaginary. When Solver evaluated the slope at $x = 0.05$ it decided to take a step downhill all the way to $x = -1.55$ which is where the value of y cannot be calculated. This is an example of a "singularity." To obviate this problem, and having understood that x must not be less than -1, a constraint has to be added to the Solver. In the "Solver Parameters" window click "Add" and enter the corresponding constraint. The complete "Solver Parameters" window and the correct coordinates of the second minimum are in Fig. 5.5.

	A	B
1	**x**	**y**
2	-1.55	#NUM!

Fig. 5.4 Oops?

[9] This result is obtained by Excel 2007. Newer versions of Excel correct some errors of this kind automatically and thereby prevent making an instructive point.

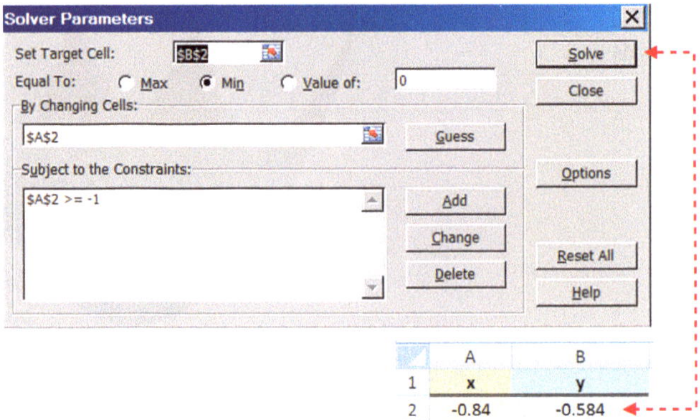

Fig. 5.5 With constraint added

The moral is that the functions for which we seek optima may present challenges to computation. The snags can materialize as division by zero, logarithms of nonpositive numbers, numbers that exceed the limits of what the spreadsheet can represent, etc.

5.3 Solver for Curve-Fitting: An Example

The main features of the Solver were presented within the context of finding the peaks and the valleys of the function in Fig. 5.1. The question is how this text-book like example relates to task of curve-fitting and parameter estimation.

To illustrate how, suppose that a parametric curve is to be fitted to the estimates of $\sigma\{\mu\}$ from Fig. 2.3. These are reproduced in columns B and D of the table in Fig. 5.6.[10] Judging by the graph on the bottom-left, $\beta_0\text{AADT}^{\beta_1}$ might be a reasonable choice function to fit; it goes through the origin and could be upward or downward bending depending on whether β_1 is larger or less than 1. If it was approximately a straight line through the data points a sensible initial guess of β_1 would be 1. This initial guess was placed in G19. Judging by the data points the ordinate of such a straight line at AADT = 10,000 would be about 3, making the initial guess of β_0 into 0.0003. However, to test the robustness of SOLVER and to illustrate the value of the "correspondence line" the poor initial guess of 0.0010 was placed into G18 instead. Using these initial guesses the formula $\beta_0\text{AADT}^{\beta_1}$ was coded into column E and the spreadsheet computed the tentative Fitted Values.

[10] To download this spreadsheet go to http://extras.springer.com and enter the ISBN of this book. Look in the "Spreadsheets" folder for "Chap. 5. Fitting a curve to estimates of sigma mu.xls or xlsx."

Average AADT	Data I&F accidents	$\hat{\sigma}\{\mu\}$	Fitted Values $\beta_0 AADT^{\beta_1}$	Computations Weights ∝ Accidents	Weighted squared differences (SD)
500	376	0.38	0.50	0.138	0.002
1500	445	0.64	1.50	0.163	0.121
2500	382	1.15	2.50	0.140	0.253
3500	381	1.53	3.50	0.140	0.544
4500	421	2.47	4.50	0.154	0.637
5500	176	1.96	5.50	0.064	0.808
6500	118	1.93	6.50	0.043	0.903
7500	151	3.14	7.50	0.055	1.052
8500	83	2.97	8.50	0.030	0.929
9500	102	5.46	9.50	0.037	0.610
10500	81	2.30	10.50	0.030	1.994
11500	13	3.45	11.50	0.005	0.309
Sums	2729	27.39	72.00	SSD=	8.162
				β_0=	0.0010
				β_1=	1.000

Target

By Changing

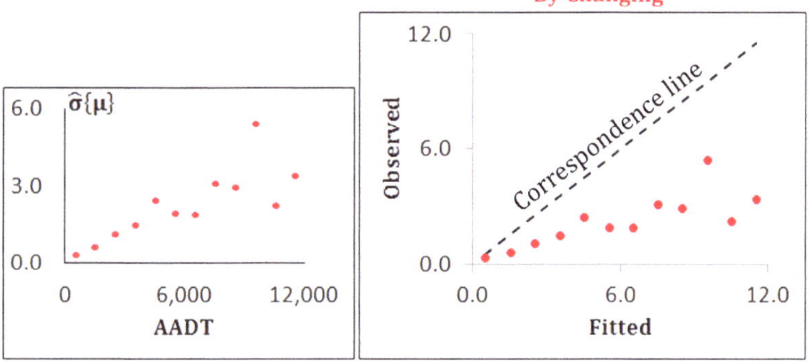

Fig. 5.6 Curve-fitting spreadsheet for $\sigma\{\mu\}$ as a function of AADT

The correspondence graph on the bottom right shows that with the chosen initial guesses the fitted values are too large. The same can be gleaned from the discrepant sums in columns D and E.

A time honored curve-fitting strategy is to minimize the sum of squared differences (SSD) between the data points and the fitted values[11]. Thus, for example, at AADT = 500 the squared difference (SD) between the fitted value and $\hat{\sigma}\{\mu\}$ is $(0.50–0.38)^2$. When calculating the SSD it would be inappropriate to give the same weight to an estimate of $\sigma\{\mu\}$ that is based on 376 accidents (cell C5) and to one based on 13 accidents (cell C16). Assuming that the weights are proportional to the number of accidents on which $\hat{\sigma}\{\mu\}$ are based[12] the weight in F5, e.g., is

[11] These differences will also also be called "residuals."

[12] Ideally the weight should be inversely proportional to the variance of the data point.

376/2,729 = 0.138. Using the weights in column F, the weighted squared differences are in column G and the weighted SSD is in G17.

The role of the Solver is to find values of β_0 and β_1 (in the "by changing" cells G18:G19) that will make the weighted SSD (in the "Target Cell" G17) minimum. Clicking on the "Data" tab, then on "Solver," preparing the "Solver Parameters" window as in Fig. 5.7, and clicking on "Solve" produced the Solver results shown.

The estimate of β_0 (0.0023) is of no interest; it is just a scaling multiplier. The estimate of β_1 (0.805) gives the model equation its shape. It seems that $\sigma\{\mu\}$ increases less than linearly with AADT.

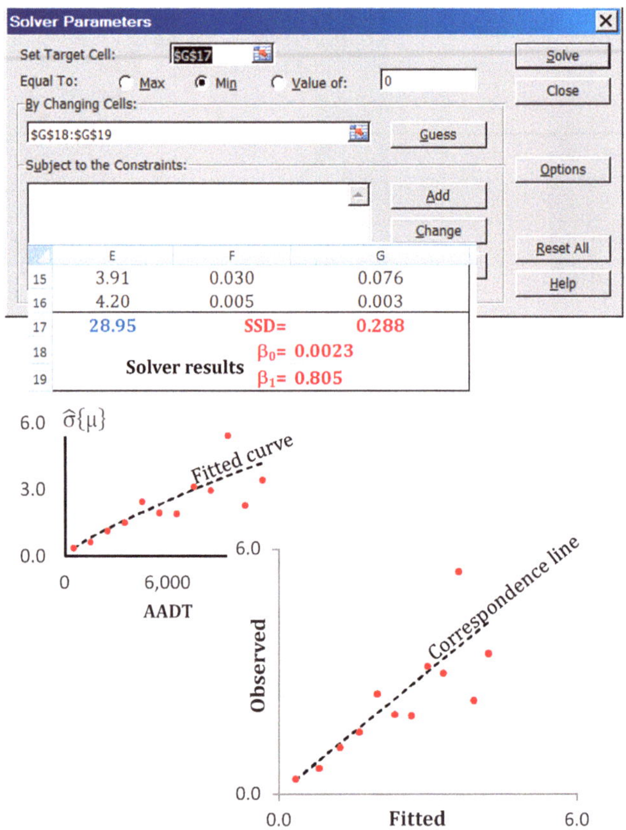

Fig. 5.7 The "Solver Parameters" window, Solver results, and the Correspondence graph

5.4 Initial Guess and Parameter Scaling

For the successful use of Solver two issues require attention. The first is that of choosing reasonable initial guesses for model parameters; the second is guarding against potential failure of the Solver algorithm.

Note that in Fig. 5.6 the sum of fitted values is 72.00 which is about three times the sum of the $\hat{\sigma}\{\mu\}$'s, 27.39. This is a sign of poorly chosen initial guesses for β_0 and/or β_1. A useful tool choosing the initial values of parameters is the "Correspondence Graph" at the bottom right of Figs. 5.6 and 5.7. The abscissae of the circles are the fitted values and their ordinates are the observed values, $\hat{\sigma}\{\mu\}$ in this case. The dashed "correspondence line," the diagonal, has the fitted values as both abscissae and ordinates. For a good initial guess the circles should hug the correspondence line. The tighter the hug the better is the correspondence between what was observed and what the model equation predicts.

The second issue is the difficulty which the Solver algorithm may encounter when the parameter values are very disparate. Thus, for example, in Fig. 5.6 the initial guesses of 0.001 and 1 differ by three orders of magnitude. In such cases the round-off errors may build up to the point where the Solver can no longer reliably find the optimal solution. To guard against this potential snag an easy remedy is to check the "Use Automatic Scaling" box in the "Options" command button. This is not always effective. A more reliable option is to rescale the variables in the model equation so that all parameters are within not more than three orders of magnitude of each other. Thus, for example, one can replace Average AADT in the model equation by, say, AADT/1,000 and thereby make the magnitude of β_0 similar[13] to that of β_1.

In sum, when the Solver is used to fit curves and estimate parameters, the following are the main steps:

1. Choose the model equation to be fitted. (Here it was $\beta_0\text{AADT}^{\beta_1}$.)
2. Input into a range of cells that can be later conveniently (contiguously) selected some good initial guesses for the parameters. (Here the range was G18:G19.)
3. Input the formula that computes the fitted values. (Here it is in column F.)
4. Optionally add a correspondence graph.
5. Decide on the criterion by which to judge the goodness of a fit. It is usually either (a) some function of the differences between the observed and the fitted values or (b) the likelihood of the observed values in the light of the fitted ones. (The criterion here was the sum of weighted squared differences.)
6. Use the Solver to find the parameters which make for the best fit.

With this we have the tool for estimating the parameters of a "parametric" SPF.

[13] Now, at optimum, $\beta_0 = 0.59$.

5.5 Summary

Because the safety of units depends on several traits, when bins are narrow enough to make the SPF practically useful, the data becomes too sparse for estimates to be sufficiently accurate. As a result, parametric curve-fitting is unavoidable; the modeler has to specify a model equation and estimate its parameters. Parameter estimation requires optimization. For study, for instruction, and for practical road safety modeling one can use an optimization tool available in Excel: the Solver.

A simple text-book like example was used to explain how to use the Solver and how it finds minima and maxima. The use of Solver for curve-fitting was explained by an example. The task was to estimate the parameters of a model equation linking AADT and estimates of $\sigma\{\mu\}$ obtained earlier in Chap. 2. At the center of the process were the "fitted values," one for each AADT that has an estimate of $\sigma\{\mu\}$ associated with it. These fitted values were computed by the spreadsheet formula that represented the chosen model equation using an initial guess for the unknown parameters. The Solver was then asked to find those values of the parameters that made the weighted sum of squared differences between the fitted values and the estimates of $\sigma\{\mu\}$ minimum.

While the Solver is a robust tool there are two main problems which may cause it to fail. First, while the Excel Solver is good at converging on local optima, finding global minima or maxima is a challenge. The Solver can fail to converge on the global optimum because it is trapped in a local optimum which lies between the initial guess and the global optimum. The closer is the initial guess to the global optimum the lesser is the chance of encountering such a blocking local optimum. The "correspondence graph" can help with the identification of a good initial guess. Second, the Solver algorithm may fail for computational reasons when the parameters to be estimated differ by several orders of magnitude. This problem may be obviated by appropriately scaling the model variables.

References

Fylstra D, Lasdon L, Watson J, Waren A (1998) Design and use of the Microsoft Excel Solver. Interfaces 28(5):29–55

Harmon M (2011) Step-by-step optimization with Excel solver. The Excel Statistical Master. http://excelmasterseries.com/D-_Loads/New_Manuals/Step-By-Step_Optimization_S.pdf

Weise T (2008) Global optimization algorithms – theory and application http://www.it-weise.de/projects/book.pdf

A First Parametric SPF

6

Abstract

The process of SPF development favors the gradual buildup of the model equation. In this chapter Segment Length is the first variable placed into a simple model equation and its parameters are estimated by minimizing the sum of (weighted) squared differences on a curve-fitting spreadsheet. Two general questions arise: (1) How accurate are the parameter estimates and (2) whether an SPF can be used to predict the safety effect of design changes and interventions.

6.1 The Approach to Parametric SPF Modeling

A parametric SPF is a mathematical function of traits (variables) and of parameters. The activity of fitting a parametric SPF to data alternates between choosing the variables from which the model equation is to be made, determining the form of the function (i.e., how the variables and parameters should combine into an equation), estimating the value of the parameters, and examining the "goodness" of the fit. These cyclical activities are shown schematically in Fig. 6.1.

The task of SPF development is difficult for several reasons. First, the regularity which is hidden in the data and which we seek to capture by a mathematical expression is some kind of multidimensional surface that is not easily visualized. We can only get two-dimensional glimpses of it by the initial EDA. Because lurking variables and confounding are omnipresent dangers, what we can see in graphs may

E. Hauer, *The Art of Regression Modeling in Road Safety*,
DOI 10.1007/978-3-319-12529-9_6

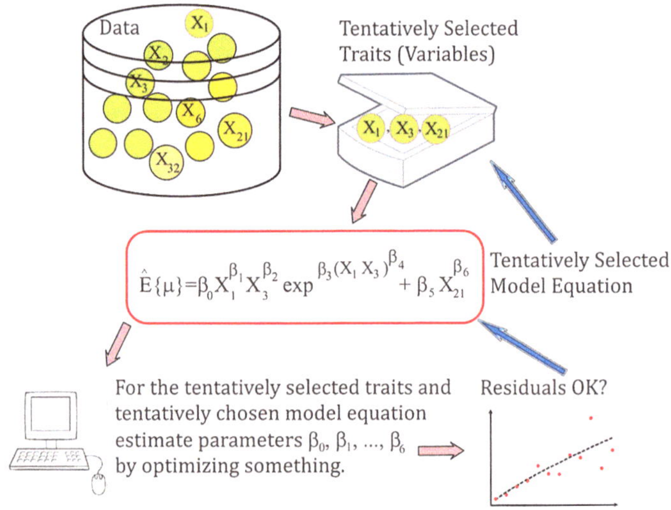

Fig. 6.1 Parametric SPF development

not reflect the workings of what is shown along the axes but of different causes.[1] Second, the main features of the hidden regularity are presently not known by some logic-based reasoning or data-based theories. By logic one can only assert that on segments of zero length and on segments with no traffic there will be no accidents. That means that at the origin the chosen model equation should have a zero intercept.[2] All other assertions about how the expected number of accidents might change as a function of traits are open to challenge. Third, we know that there are safety-related traits for which there are no data and that these are correlated to variables for which data are available. Thus, for example, rich people tend to drive safer cars, use better roads, and be safer drivers than low income persons. But in SPF modeling only the traits of the road are commonly used as variables while road user income and education do not feature in the models. If so, a part of what the SPF model will attribute to the road may reflect affluence. That is, the variables that are present in the SPF are to an unknown extent proxies for variables not present in the SPF. Fourth, the data we have is often inaccurate and

[1] This is commonly attributed to "lurking variables." A lurking variable is one that is not in the model equation but affects one or more predictor variables and the dependent variable, thereby creating a correlation between them. But "correlation is not causation." To illustrate, consider the relationship between the shoe sizes of elementary school students and their scores on a standard reading test. While these are correlated, thinking that larger shoe size causes higher reading scores is as absurd as thinking that high reading scores cause larger shoe sizes. Here the lurking variable is age. As the child gets older, both their shoe size and reading ability increase.

[2] Even this logical requirement is more of esthetic significance than of practical importance. Very short road segments with very little traffic are of no interest in applications. Therefore, should it turn out that a non-zero intercept improves the fit, the esthetic price may be worth paying.

their causal meaning is diminished by averaging. To illustrate, estimates of AADT used in modeling are usually "factored up" from counts over few summer days conducted once in 3–4 years; these estimates are hardly accurate. Besides, while there may be a cause–effect relationship between hourly accidents and hourly traffic flows, the causal linkage between the daily average traffic over a year and annual accidents is certain to be causally remote.[3]

The main difficulties:

- What the EDA can show is subject to much confounding
- There is no theory to guide the choice of the model equation
- Variables in the model equation are proxies for variables not in the model
- Data are inaccurate and aggregated

At the heart of parametric curve-fitting is the model equation. As noted earlier, even when a model equation is not used, curve-fitting is associated with several problems: it takes away some of what is unique at a certain level of a trait, it creates features of the fitted curve that are essentially foreign and peculiar when judged against local data, and when data is sufficient the fitting may do more harm than good. The need to have a model equation adds another major concern. Now the curve or the surface is not free to respond to the pull of the data. Being restricted to some functional family of simple mathematical expressions, the model equation is constricted to a prespecified shape. An increasing function cannot have a maximum, a function with one inflection point cannot have two, a continuous function cannot accommodate jumps, etc. The model equation is usually built up by the addition or multiplication of simple functions of one variable. Seldom are addition and multiplication mixed and only rarely do the building blocks depend on two or more variables. However, there is no reason why the regularity we seek to replicate and the causes of which are many and complex, should follow any simple algebraic expressions. In short, by having to conform to an agglomeration of simple algebraic expressions, a good parametric curve fit can still be substantially worse than a nonparametric one. Even when masterfully executed, the parametric fit will suffer from all the shortcoming of a nonparametric one and then some.

A model equation puts a straightjacket on the data. A well-fitting straight-jacket is an oxymoron.

In the chapters to follow the parametric SPF will be built from the ground up. Variables (traits) will be introduced into the model equation one after another, pausing after each addition to make sure that that every interim SPF is as good as it

[3] See, e.g., Mensah and Hauer (1998).

can be.[4] The advantages of such an incremental buildup are several. First, it suites the didactic purpose of the book; concepts, considerations, and tools can be introduced at the point at which they are needed. Second, gradualism is of essence. The insertion of every new variable into the model equation will be preceded by a special purpose EDA that will inform the modeler about whether the variable is worth introducing and, if yes, what mathematical function might best represent it. Also, once a new variable is introduced into the model equation, we will examine whether the fit is satisfactory, whether there are outliers to be discarded, and how the fit could be further improved. In this manner the model equation is allowed to emerge gradually and need not be postulated in its fully loaded form right from the start. The third advantage is practical – at the end of each stage there will be a usable product. This is of import because the data available to practitioners are usually less extensive than the data assembled for SPF modeling. SPFs with variables for which the practitioner has no data cannot be used. To make the results of modeling useful to practitioners, one should report every SPF obtained en route to the richest and final model. In that way practitioners can use the model for which they have the data.

This chapter is about initial steps. Segment Length will be the first variable placed into the SPF and a simple model equation will represent its relationship with $E\{\mu\}$. This will facilitate the building the first curve-fitting spreadsheet and the estimation of its parameters. Parameters will be estimated by minimizing the sum of squared differences (SSD). It will be shown how $\sigma\{\mu\}$ can be estimated. All this will be the basis for elaboration in later chapters where the questions of when and how to add variables, how to chose amongst alternative objective functions, how to make use of more complex model equations, and how to examine the goodness of a fit and quality of estimation will be discussed. A separate chapter will be devoted to each topic.

Already at this initial step important questions arise. How accurate is the estimated parameter. Can one trust the usual measures of parameter accuracy? Does an SPF predict how the safety of a unit changes due to an intervention which alters the value of a variable? To what extent does the parameter of variable X_i reflect the influence of some related variable X_j which is not in the model equation? All these issues will be raised and aired.

[4] The one-after-another addition of variables is reminiscent of the "Forward Selection" approach to variable selection as opposed to "Backward Elimination" in "Stepwise Regression." The former starts with no variables in the model, adding the variable that improves the model the most, and repeating this process until no variable improves the fit significantly. The latter begins with all variables in the model, and removing those the deletion of which harms the fit least, till a parsimonious and satisfactory model remains. The hallmark of Stepwise Regression is that it is an automated process; the modeler clicks a button and the software does the rest. The modeling approach advocated here abhors automation. It rests on the belief that to build a good SPF the exercise of human insight and intelligence is essential.

6.2 A Simple Parametric SPF

In Fig. 6.1 the notation $X_1, X_2, X_3, \ldots$ was used for traits (variables) and $\beta_0, \beta_1, \beta_2, \ldots$ for parameters. This notation is useful when no specific variables (traits) are named. For named variables, it may be better to use mnemonically suggestive acronyms; AADT for average daily traffic, L for Segment Length, etc. It makes sense to begin model building with the most obvious variable, Segment Length $\equiv L$.

By what curve (function) should one represent the dependence of $E\{\mu\}$ on L? Logic suggests that when $L = 0$ then $E\{\mu\} = 0$, i.e., that the curve must start at the origin. One might also be inclined to think that the relationship is one of proportionality: twice the length – twice the accidents; that is, that $E\{\mu\} = \beta_0$L. However, in the initial EDA (Fig. 3.14) there was a hint that the function may not be a straight line and the causes of a possible nonlinearity were discussed in Sect. 3.5.

There is a multitude of functions to choose from: $\beta_0 L^{\beta_1}$, $\beta_0(L + \beta_1 L^2)$, $\beta_0 e^{\beta_1 L}$, $\beta_0 L^{\beta_1} e^{\beta_2 L}$, and many more. How then to choose one and not another? This question will be examined in detail in Chap. 10. Here, to get started, $E\{\mu\} = \beta_0 L^{\beta_1}$ will be used. For positive values of β_1, this is an increasing function that bends upward when $\beta_1 > 1$ and downward when $\beta_1 < 1$.[5]

6.3 Preparing and Using the First Curve-Fitting Spreadsheet

One of the main tools of SPF development will be the curve-fitting (C-F) spreadsheet.[6] The top and bottom parts of the first C-F spreadsheet are in Fig. 6.2. As variables will be added, the model equations grow, and other objective functions tried, the C-F spreadsheet will evolve. However, all C-F spreadsheets consist of four essential regions plus the optional "correspondence graph":

1. A region containing the data.
2. A region where the to-be-estimated parameters are.
3. A region where, using the data and parameters, the fitted values are computed.
4. A region where the objective function (e.g., SD and SSD) is computed.

The by-now familiar condensed Colorado data are in columns B and C. There are 5,323 segments – one row per segment. The initial guess at the two parameters β_0 and β_1 is in cells D2:E2; this is where the Solver begins its search. The formula

[5] A comment on the difference between β_0 and β_1 is in order. The β_1 determines the essential shape of the fitted curve. The role of β_0 is different. It merely pre-multiplies whatever expression follows and thereby makes sure that the sum of the predicted accidents is roughly the same as the sum of the observed accidents. In this sense β_0 is just a scaling factor. To fix this idea, β_0 will be referred to as the "scale parameter" and the other βs as "shape parameters."

[6] To download the spreadsheet, go to http://extras.springer.com/ and enter the ISBN of this book. Look in the "Spreadsheets" folder for "Chap. 6. First SPF, unweighted and weighted .xls or xlsx."

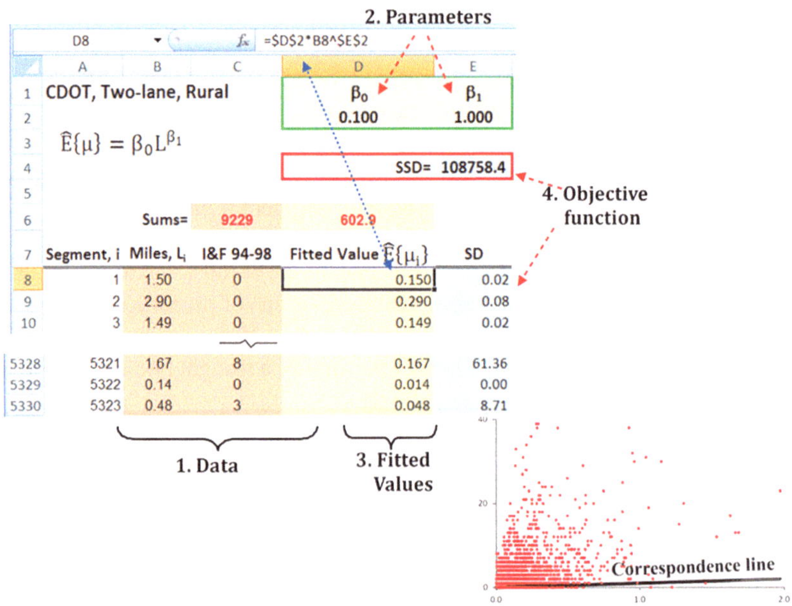

Fig. 6.2 The C-F spreadsheet and its four regions

for the model equation $\hat{E}\{\mu\} = \beta_0 L^{\beta_1}$ was entered into cell D8 and copied all the way down. In E8 is the squared difference (SD) between the accident count in C8 and the fitted value in D8. The sum of the squared differences[7] (SSD) is in E4.

That the initial guesses in Fig. 6.2 are not good is clear from comparing the sum of accident counts (9,229) to the sum of fitted values (602.9) as well as from the comparison graph. After changing the initial guess for β_0 to 1.5 the by-now familiar sequence of clicks and choices for the Solver[8] yields the fit in Fig. 6.3; parameter estimation is that simple.

[7] At this early stage of model development, the minimization of the sum of squared differences is the "objective function." The reasons for this choice are several. First, as already noted, this is a convex objective function and therefore the minimum which the Solver finds is global. For this reason, it is also a good source of initial guesses for other objective functions. Second, it is a popular choice, particularly in econometric modeling. Third, the least squares objective function has a long history going back to Carl Friedrich Gauss who is credited for developing it in 1795 (at the age of 18) and to Adrien-Marie Legendre who published it in 1806. In addition to being "time honored" this objective function has the merit of being the BLUE (best-linear-unbiased-estimator) of the parameters when observation errors have a mean of zero, are uncorrelated, and have equal variances. When the variances are not equal, as in our case, an appropriate weighting can be applied.

[8] On the "Data" tab choose "Solver," in the "Solver Parameters" window choose E4 as "Target Cell," set the radio button to "Min," choose D2:E2 as "By Changing Cells" and click "Solve."

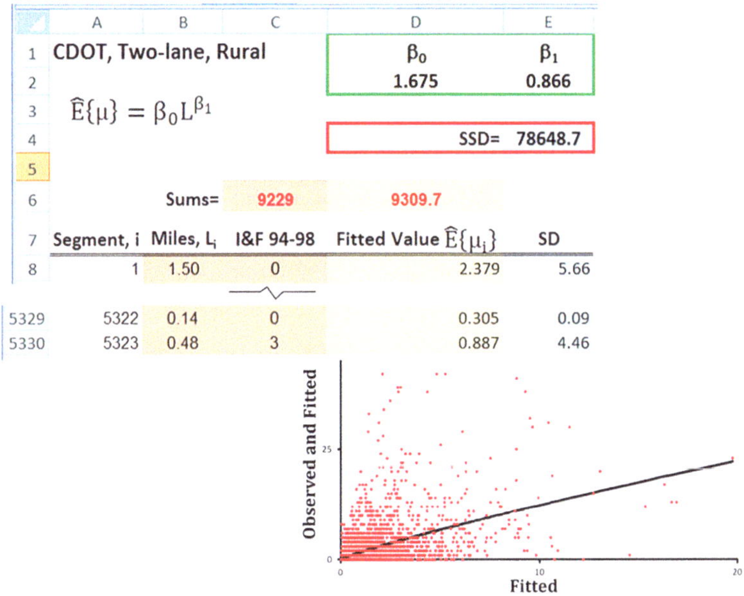

	A	B	C	D	E
1	CDOT, Two-lane, Rural			β_0	β_1
2				1.675	0.866
3	$\hat{E}\{\mu\} = \beta_0 L^{\beta_1}$				
4				SSD=	78648.7
5					
6		Sums=	9229	9309.7	
7	Segment, i	Miles, L_i	I&F 94-98	Fitted Value $\hat{E}\{\mu_i\}$	SD
8	1	1.50	0	2.379	5.66
5329	5322	0.14	0	0.305	0.09
5330	5323	0.48	3	0.887	4.46

Fig. 6.3 The solution that minimizes the sum of squared differences

Thus one part of the first parametric SPF is $\hat{E}\{\mu\} = 1.675 \times L^{0.866}$ accidents in 5 years. This estimates the average number of accidents in 1994–1998 for populations of rural road segments in Colorado of a known length. Since segments in such populations will differ in AADT and in many other traits, this is hardly a practically useful SPF. However, as will be shown, adding AADT (and other variables) to the C-F spreadsheet is straightforward; one only has to add a column with AADT data, add an AADT factor to the formula in the "Fitted value" column, add space, and initial guesses for the parameters of the new factor, and then use the Solver.

6.4 Modifying the Objective Function

This section illustrates the malleability of the C-F spreadsheet in the hands of the modeler. There is a problem with the objective function used in Fig. 6.3. Estimation by minimizing the sum of squared differences makes sense only if the variances of all observed dependent variable are the same. In road safety this is clearly not true. Some short segments had a fitted value of 0.03 while a 19.7 miles long segment had a fitted value of 22.1 accidents in 5 years. With such discrepant estimates of $E\{\mu\}$ the observed accident counts must have very different variances.[9]

When the variances of the observed values are not equal, in curve-fitting each has to be weighted in proportion to the reciprocal value of its variance.[10] In Fig. 6.4 the ratio 1/(Fitted value in Fig. 6.3) served as the raw weight[11] in column F and the sum of weighted squared differences in G4 was the "Target Cell" when Solver was used.

	A	B	C	D	E	F	G
1	CDOT, Two-lane, Rural			β_0	β_1		
2				1.666	0.859		
3	$\hat{E}\{\mu\} = \beta_0 L^{\beta_1}$						
4						wSSD=	26608.8
5							
6		Sums=	9229	9229.0			
7	Segment, i	Miles, L_i	I&F 94-98	Fitted Value $\hat{E}\{\mu_i\}$	SD	Weight	Weighted SD
8	1	1.50	0	2.360	5.57	0.420	2.34
9	2	2.90	0	4.159	17.30	0.238	4.11
10	3	1.49	0	2.347	5.51	0.423	2.33
5328	5321	1.67	8	2.589	29.28	0.383	11.22
5329	5322	0.14	0	0.308	0.09	3.274	0.31
5330	5323	0.48	3	0.887	4.47	1.127	5.03

Fig. 6.4 Using the first fitted value as variance

[9] It is commonly assumed that accident counts are Poisson distributed and Appendix A lists the supporting arguments. For the Poisson distribution, the Mean = Variance. It follows that the accident counts on segments with different means have different variances.

[10] When a curve is being fitted to data, it is as if each observed value was pulling the to-be-fitted curve towards itself; the larger the variance of the observed value, the lesser is the force with which it pulls. Here is why. Call X_i the observed value for unit i and call its variance V_i. Assume that for units 1 and 2 $V_1 > V_2$. Imagine that X_2 is the average of n independent observed values the variance of which is V_1. The question is how many observed values with variance V_1 must be averaged so that the average has a variance V_2? Answer: Because the average is $\frac{\Sigma_1^n X_i}{n}$ the variance of the average is $\frac{nV_1}{n^2} = \frac{V_1}{n}$. If $\frac{V_1}{n}$ is to equal V_2, it is as if X_2 was the average of $n = \frac{V_1}{V_2}$ observed values with variance V_1. That implies that when summing residuals the weight for unit 2 should be V_1/V_2, i.e., in proportion to the reciprocal values of its variance. An early reference is Aitken (1935).

[11] One could normalize the raw weights to make their sum equal 1 but, since this amounts only to multiplying the sum of weighted squared deviations by a constant, doing so would not affect the minimization.

Thus, one part of the first parametric SPF is:

$$\hat{E}\{\mu\} = 1.666 \times (\text{Segment Length})^{0.859} \text{ accidents in 5 years} \quad (6.1)$$

Between the unweighted and the weighted C-F the estimate of β_1 changed but little, from 0.866 to 0.859. Note however that the weights in Fig. 6.4 were determined by the fitted values from the earlier spreadsheet (Fig. 6.3) and these have now changed. Allowing the weights to adapt to the current value of the parameters[12] is simple; the formula in column F is changed to 1/(the corresponding fitted value in column D). With this change the weights change as the Solver seeks the optimum. After convergence, the estimates of β_0 and of β_1 were 3.452 and 0.737.

The role of SPFs is to produce estimates of $E\{\mu\}$ and $\sigma\{\mu\}$. So far the discussion revolved around the estimation of $E\{\mu\}$. In the next section the focus is on the estimation of $\sigma\{\mu\}$.

6.5 Estimating $\sigma\{\mu\}$

It can be shown[13] that for a population of units with diverse μs in which each unit is associated with one accident count

$$V\{\mu\} = \text{Variance of accident counts} - \text{Mean accident count} \quad (6.2)$$

In a C-F spreadsheet each row is a sample of one from an imagined population. Thus, for example, in Fig. 6.4 there are 5,323 imagined populations, each segment being a sample of one with an unknown μ and an accident count. To use Eq. (6.2) in this circumstance, for each imagined population (row) one may estimate the variance of accident counts by the squared difference between the observed accident count and the fitted value (the SD) and the mean accident count by the fitted value. If so,

$$\hat{V}\{\mu\} = \text{SD} - \text{Fitted value} \quad (6.3)$$

These estimates can be computed for every segment using the values in columns C and D of Fig. 6.4 and are shown by the cloud of dots in Fig. 6.5.[14] The solid curve is the Nadaraya-Watson nonparametric fit to these points.[15] Once again a pattern masked by randomness is revealed by the magic of nonparametric regression.

[12] This procedure goes under the name of "Iteratively reweighted least squares."

[13] See Appendix C.

[14] When estimating for one population and the difference is negative, it is best to use $\hat{V}\{\mu\} = 0$. However, when estimating for many populations doing so would entail a systematic bias. To avoid this bias, the negative estimates were retained. The striated bands of points come from the use of the integers 0, 1, ... in computing the SD.

[15] The bandwidth here was 2.

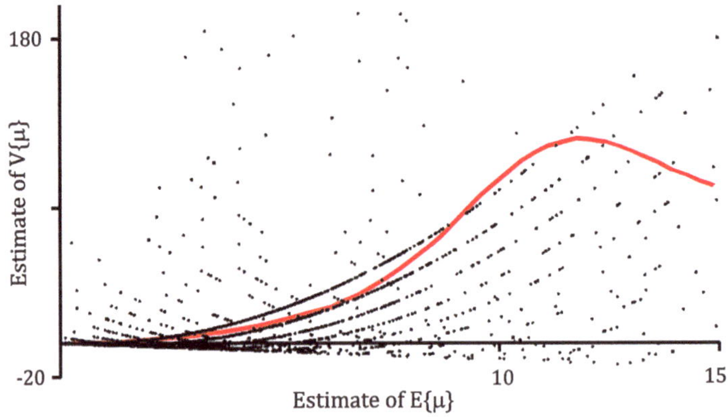

Fig. 6.5 How $\hat{V}\{\mu\}$ depends on $\hat{E}\{\mu\}$

Using the N-W data points in the range shown in Fig. 6.5 a weighted least squares regression yields:

$$\hat{V}\{\mu\} = 0.744\hat{E}\{\mu\}^{2.00} \tag{6.4}$$

Equations (6.1) and (6.4) are the products which an SPF is to provide: estimates of $E\{\mu\}$ and of $\sigma\{\mu\}$.

The main aim of this chapter was to create a first SPF, a point of departure for the development and enrichment to come. Already at this initial step some important questions arise. The first question is about the accuracy[16] of parameter estimates. Were the estimate of β_1 close to 1 one might conclude that the expected number of accidents is approximately proportional to Segment Length. However, since values less than 1 were obtained it is possible that the relationship is not one of proportionality. How trustworthy is this early indication of nonlinearity? How accurate is the estimate of β_1?

The second question is about the meaning and use of parameter estimates. To illustrate, suppose that a highway designer can choose between two alternative horizontal curve designs; one with a 1.0 mile long tangent and the other with a 2.0 mile tangent. By Eq. (6.1) the first tangent is expected to have 1.666 fatal and injury accidents in 5 years while on second less than 2×1.666 accidents are expected.[17] Can it really be that on otherwise identical roads a segment which is twice as long is not expected to have twice as many accidents? If this cannot be true, then can SPFs be used to make design choices or to predict the effect on $E\{\mu\}$ of change in the value of a predictor variable? These issues will be examined in the next two sections.

[16] Accuracy is the degree of closeness between the measurements of a quantity and its true value (JCGM 2008).

[17] $2^{0.859} = 1.81$.

6.6 The Accuracy of Parameter Estimates

It is the very nature of statistical estimates that they differ to an unknown extent from what they are estimating – the estimand. So far, using the same data and model equation three estimates of β_1 were obtained: 0.866, 0.859, and 0.737. This in itself is an indication that parameter estimates will vary and are to some extent inaccurate and uncertain. The sources of uncertainty are many and can be ascribed to two sources: (1) those rooted in the modeling process – the "modeling inaccuracy," and (2) those deriving from uncertainty in the data – the "statistical inaccuracy."

There are two sources of uncertainty surrounding parameter estimates:
1. Modeling inaccuracy
2. Statistical inaccuracy

As the SPF will evolve from chapter to chapter, and as modeling choices will be made, parameter estimates will be seen to change. They will change when a new variable is introduced into the model equation, they will depend on what functions are chosen to represent variables, and they will be found to vary with the choice of the objective function. These sources of parameter estimate variability derive from the availability of data and from the choices the modeler makes; they will be referred to as the "modeling inaccuracy."

There is, of course, also the more commonly recognized source of inaccuracy surrounding parameter estimates, that which derives from the uncertainties inherent in the data used. At this point in model development, with Segment Length the only variable, only the accident counts are a source of data-related uncertainty. To explain, if the same period could be repeated, and if nothing changed about the safety-related traits of the Colorado road segments, a different set of accident counts would materialize. This is the same kind of variation in outcomes as that between the successive casts of a die or tosses of a coin. With a different set of accident counts, a different set of parameter estimates would obtain. That part of statistical inaccuracy surrounding the estimate of β_1 which is due to the randomness of accident counts will be examined below. As additional variables will be added to the model equation, new sources of data-based uncertainties will emerge. Thus, for example, estimates of AADT are subject to considerable uncertainty. After AADT and other variables will be added to the SPF the statistical inaccuracy issue will be revisited.[18]

[18] Estimates of the Annual Average Daily Traffic are usually produced by "factoring up" counts conducted over a few hours on few summer days. Such counts are conducted once every few years and AADT estimates for the gap years are based on interpolation. How to account for such "errors in variables" will be discussed in Chap. 11.

6.6.1 The Statistical Inaccuracy of β_1

The path from data to parameter estimates is through computation. In this chapter computation was of fitted values, of weighted squared differences, and of minimization using the Solver. Whatever the details of these computations, in the final account the parameter estimates are made out of the data; were the data different so would be the parameter estimates.

To explain more vividly, imagine that each data point (ordinate = Segment Length, abscissa = Accident Count) is tied to the to-be-fitted curve by a vertical rubber band. When not stretched, the rubber bands are of equal length but differ in elasticity. When stretched, the pull of on the curve is proportional to the square of the length of the rubber band and on its elasticity; in this metaphoric picture elasticity represents "weight" in the weighted least squares regression. The position of the fitted curve, and thus the magnitude of β_1, is where the rubber band forces are in equilibrium. Consider now a data point which is above the fitted curve. If due to the randomness inherent in accident counts the point was higher, its upward pull on the curve would be stronger, the curve would move a bit upward, and therefore the estimate of β_1 would be slightly larger. Conversely, if due to the same randomness in accident counts the data point was lower than it is now, the band would pull with less force, the curve would shift a bit downward and the estimate of β_1 would decrease. In short, as the position of the data points varies due to randomness so will the estimate of β_1. Here the question is how much uncertainty in the estimate of β_1 is caused by the randomness inherent accident counts. How different the estimates of β_1 would be if the observed accident counts were different can be ascertained in various ways.

To explain further Fig. 6.6 shows how the weighted SSD[19] changes as a function of β_1. Based on the results in Fig. 6.4, the minimum is at 0.859. It is characteristic of extrema of smooth functions that they are surrounded by a nearly flat region. The width of that plateau indicates how sharply the minimum is defined. The question is by how much would the position of the minimum change due to random changes in the accident counts.

When estimation is by maximizing likelihood, the customary method for assessing the accuracy of parameter estimates is by computing the inverse of the observed Fisher Information Matrix.[20] In the spreadsheet environment is just as easy and perhaps more appropriate to do so by Monte Carlo simulation.[21] The idea is to generate the accident counts for each road segment from a Poisson distribution such that the μ of the segments comes from a Gamma distribution for which the central moments are those given by Eqs. (6.1) and (6.4). Applying the Solver to the newly

[19] 26,600 has been subtracted from the ordinate.

[20] This obscure sounding phrase will be explained and applied once the likelihood concept is introduced in Chap. 8.

[21] The advantage of estimating parameter accuracy using Monte Carlo simulation is that it can be used even when parameter estimation is not by maximizing likelihood and when there are errors in variables (e.g., in AADT). For detail, see Straume and Johnson (2010).

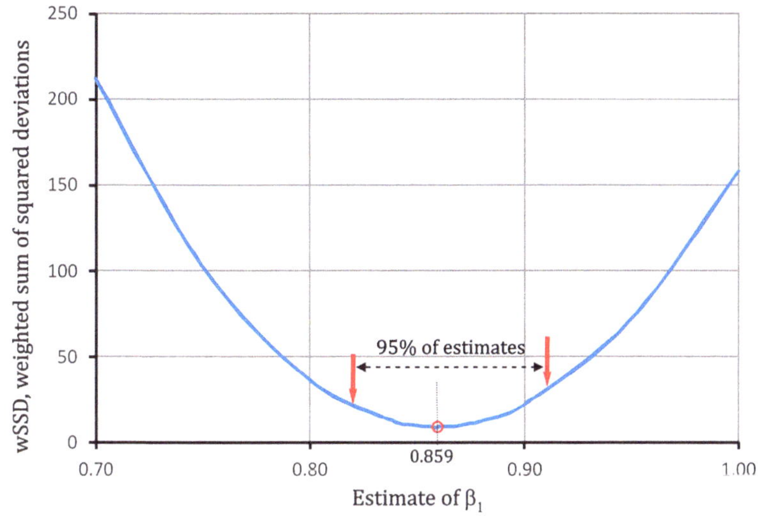

Fig. 6.6 The SSD "Valley"

generated set of accident counts, a new estimate of β_1 is obtained. Repeating this process 100 times[22], the standard error of the estimate of β_1 was ± 0.024 and 95 % of the estimates were between 0.82 and 0.91 as shown in Fig. 6.6.

6.6.2 The Incompleteness of "Statistical Inaccuracy"

It is common practice to describe the inaccuracy of parameter estimates by standard errors, confidence intervals, *t*-statistics, and *p*-values.[23] These measure only the "statistical inaccuracy." And yet, the uncertainty surrounding parameter estimates consists of both the "statistical" and the "modeling" inaccuracy. The neglect of modeling inaccuracy in the reporting of the reliability of parameters requires attention.

Modeling inaccuracy is difficult to measure. In consequence, it often remains unconsidered and unmentioned. One way to think about the implications of this omission is to list the conditions under which the consideration of only the statistical inaccuracy would suffice. The main conditions are three: (a) there are no variables which are presently not in the model equation and the inclusion of

[22] To download this spreadsheet, go to http://extras.springer.com/ and enter the ISBN of this book. Look in the "Spreadsheets" folder for "Chap. 6. Simulation to determine parameter accuracy. xls or xlsm."

[23] The *t*-statistic is (usually) the estimate of the parameter divided by its standard error. The *p*-value is the probability of obtaining a test statistic that is at least as extreme as what was observed when the null hypothesis is true. Thus, e.g., if the null hypothesis in $E\{\mu\} = \beta_0 L^{\beta_1}$ is that $\beta_1 = 1$, then obtaining $p < 0.01$ is a strong presumption that the $E\{\mu\}$ is not proportional to L.

which would alter the parameter estimates now in the model; (b) the assumed model equation for $E\{\mu\}$ and $\sigma\{\mu\}$ is that which really generated the data; (c) the objective function used to estimate the parameters is the only sensible choice.

Are these conditions met for the SPF of Fig. 6.4? Condition (a) is obviously untrue; at this stage not even AADT is in the model equation. Condition (b) is not met because, as noted earlier, there is no logical or empirical reason why the relationship between expected accidents and segment length should follow any simple mathematical expression, let alone the specific function in Eq. (6.1). Condition (c) is not fulfilled because parameters can be estimated not only by minimizing the sum of weighted differences but also, say, by maximizing likelihood or by minimizing the sum of absolute differences; each objective function yielding a different estimate of β_1. One cannot say with conviction which objective function is the right one.

In sum, in the present SPF (and also in general) all three assumptions are untrue to some degree. How large is the resulting modeling inaccuracy cannot be said but parameter estimate inaccuracy is always larger than what is reported. As our SPF will evolve, as new variables will be added, and as choices of functions and alternative ways of parameters estimation will be examined, the change in previously estimated parameters will become obvious and the consequences of deviation from assumptions (a), (b), and (c) will become clearer.

At the end of Sect. 6.5 two questions were raised. The first, about the accuracy of parameter estimates, was dealt with in this section. The second, about whether SPFs can be used to predict the safety effect of design choices and interventions, will be discussed next.

6.7 Regression, Design Choices, Interventions, and Safety Effect

Early on the distinction was made between SPFs for applications and those focused on research.[24] When the focus is on applications, the purpose of the SPF is to produce estimates of $E\{\mu\}$ and $\sigma\{\mu\}$ for use in practice. When the focus is on research, the aim of the SPF is to predict the effect of cause, to predict the safety effect of design choices or interventions, and to be the source of Crash Modification Factors and Functions. I do not think that regressions based on cross-sectional data can be trusted to predict the effect of cause. In this section I explain why.

For the parametric SPF in this chapter, Segment Length was the only variable, the function $\beta_0 L^{\beta_1}$ was chosen to represent its relationship with $E\{\mu\}$, data about 5,323 Colorado road segments were used to estimate the two parameters values, and minimization of wSSD produced the parameter estimates 1.666 and 0.859. These first steps already provide a sufficient and tangible setting for examining the causality-in-SPFs question. Can the equation $\hat{E}\{\mu\} = 1.666L^{0.859}$ be used, as one would any mathematical expression, to compute what effect some change in L has on $E\{\mu\}$?

[24] See Sect. 1.5.

To many the very asking of this question may seem odd and the answer obvious. Witness the vast number of peer-reviewed articles in the road safety literature in which the possibility that a regression equation may not be used to predict the safety effect of design choices and of interventions simply does not arise; causal interpretations of regression equations are offered as a matter of course, without caution or qualification. Perhaps this is so because from early on we are taught to use mathematical equations in a manner that suggests cause and effect: change the value of a variable on the right-hand side of an equation and you can see how doing so changed the value on its left-hand side. Perhaps the lack of questioning comes from later education in which logical and empirical laws presented to us in the form of mathematical equations reinforcing the same belief. Thus, for example, we are taught how to calculate the change in pressure of an ideal gas as a result of a change in its temperature, volume, or the amount of gas. Alternatively, moving variables from one side of the equation to the other, how temperature will change as a result of manipulating pressure.

There is, however, a distinction to be made between the two kinds of otherwise identical looking expressions and it was clearly stated by Simon (1953): *When we say that A causes B, we do not say that B causes A; but when we say the A and B are functionally related (or interdependent), we can equally well say that B and A are functionally related (or interdependent). Even when we say that A is the independent variable in an equation, while B is the dependent variable it is often our feeling that we are merely stating a convention of notation and that, by rewriting our equation, we could with equal propriety reverse the roles of A and B.* (p. 51)

Failure to distinguish between mathematical expressions that convey cause and expression that describe functional relationships makes it difficult to accept the caution that, unless some difficult-to-meet (and not currently agreed upon) conditions are satisfied, in spite of similarity in appearance, a fitted regression model equation may not be used in the same way as the familiar mathematical equations of functions. More specifically, that the fitted model equations should not be trusted to predict the safety effect of interventions and are not a reliable source of Crash Modification Functions (CMFs). Two specific examples will serve for illustration. These will be followed by a general discussion of this vexed issue.

6.7.1 A Road Design Example

Suppose that horizontal curves A and B in Fig. 6.7 are two road design alternatives. If curve A is chosen, the tangent is 1 mile long while with curve B it is 0.5 miles long. The designer asks how many accidents to expect on the tangents of curves A and B assuming that there is no change in any safety-related trait along the entire 1 mile long tangent of curve A.

Doing the computation with $\hat{E}\{\mu\} = 1.666L^{0.859}$, one expects 1.67 accidents in 5 years on the 1 mile long tangent of curve A and 0.92 accidents on the 0.5 miles long tangent of curve B. From this the designer might conclude that choosing

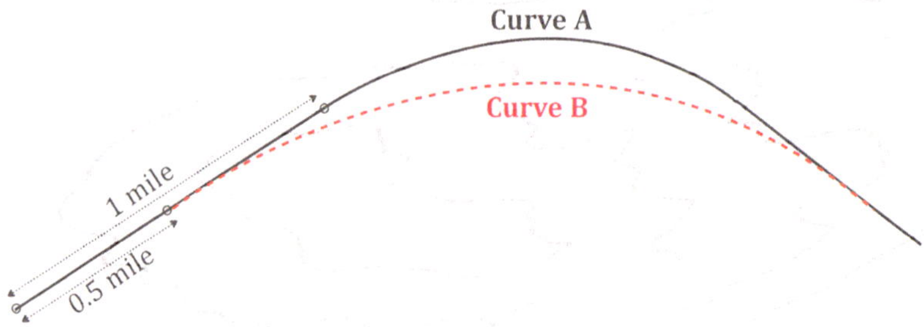

Fig. 6.7 Two horizontal curve design alternatives

design B instead of A would save on the tangent part leading into the curves 1.67−0.92 = 0.75 accidents in 5 years. The logical difficulty is that the 0.5 miles eliminated by choosing design B was said to be identical in all safety-related traits to the 0.5 miles common to both tangents where one expects 0.92 accidents in 5 years. Why should one expect on the eliminated 0.5 miles fewer accidents than on the identical 0.5 miles remaining? The question is not which of the two numbers 0.75 or 0.92 is right or more appropriate; the question is why what looks like an ordinary mathematical equation leads to a logical quandary?

The answer is this: Recall that the regression equation was fitted to cross-sectional data[25] about 5,323 road segments of varying lengths. Using such data one can justly state how many accidents are expected on segments found to be 0.5 miles long, and how many on other segments found to be 1 mile long. However, one must allow for the possibility that segments found to be about 0.5 miles long may not have the same safety-related traits as segments found to be 1 mile long. Therefore, based on cross-sectional data, one cannot say how many accidents to expect on a 1 mile long segment which has the same safety-related traits as are found on 0.5 mile long segments.

As noted earlier, there are reasons why one segment in the database is short and another is long. The reasons may have something to do with the density of major intersections, type of terrain, roadside development, history of road building, administrative boundaries, etc.[26] Each of these reasons can be associated with various safety-related traits. Thus, for example, longer distances between major

[25] Cross-sectional data describe the traits of a collection of units without there being a specific intervention or treatment applied to them at a certain time.

[26] Thus, e.g., in the Colorado data segments that are 0.4–0.6 miles long have an average AADT of 2,899 vpd and 31 % are in mountainous terrain while for segments that are 0.9–1.1 miles long the average AADT is 1,767 and only 5 % are in mountainous terrain.

intersections may be associated with less traffic, poorer road users, tree-lined or steep roadsides, more animals crossing the road, etc. Some such traits are known (e.g., AADT) and their influence on $\hat{E}\{\mu\}$ can perhaps be accounted for, while other traits are not in the database (e.g., road user income or frequency of animal crossing) and cannot be accounted for. It follows that $\hat{E}\{\mu|L = 0.5 \text{ miles}\}$ reflects the influence of one set of safety-related traits while the $\hat{E}\{\mu|L = 1 \text{ mile}\}$ may reflect a set of different safety-related traits. Therefore the two $\hat{E}\{\mu\}'s$ are not for segments that are identical in all traits except for length. But the designer asked "how many accidents to expect on the tangents of curves A and B assuming that there is no change in any safety-related trait along the 1 mile long tangent of curve A." This question cannot be answered by using $\hat{E}\{\mu|L = 0.5\}$ and $\hat{E}\{\mu|L = 1\}$ because they pertain to segments with different safety-related traits.

In fact, were the safety-related traits of the 0.5 mile and 1 mile long tangents identical, as the designer was assuming, then, by the same logic by which $1 + 1 = 2$, it must be true[27] that $E\{\mu|L = 1\} = 2 \times E\{\mu|L = 0.5\}$. For this equality to hold (when $\beta_0 L^{\beta_1}$ is the model equation), it is necessary that $\beta_1 = 1$. That the estimate turned out to be 0.859 is a sign that at this stage of modeling, β_1 reflects also the influence of various known and unknown safety-related traits other then Segment Length (L). If so, the model equation with $\beta_1 = 0.859$ may not be used to compute the effect on $E\{\mu\}$ of changes in L alone. Conversely, should at a later stage in modeling the estimate of β_1 become indistinguishable from 1 that would be a sign that L reflects only the influence of Segment Length and that the influence of other safety-related variables has been satisfactorily accounted for.

6.7.2 A Speed-and-Safety Example

To provide another cautionary example in a different setting consider Fig. 6.8 based on data assembled by Garber and Gadirau (1988) for freeway, arterial and collector road segments. Each square is an estimate of the mean accident rate of a population of road segments having approximately the same average speed.

It does appear that there is a regular relationship between accident rate and average speed and that a smooth function could be fitted to these points. But road segments with an average speed of about 45 mph (population A) are not like those in population B where the average speed was about 60 mph. The difference between the mean accident rates of these two populations reflects all these differences, not only the difference in mean speed. It would therefore be nonsensical to think that a function fitted to these points can predict how the accident rate of a road segment

[27] Recall that in Sect. 1.2 the axiom was that units identical in all safety-related traits have the same μ.

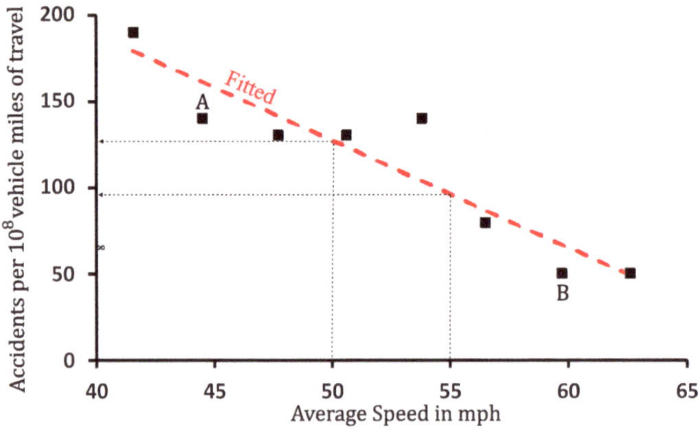

Fig. 6.8 How the accident rate declines with average speed

would change if all its traits remained the same and only the average speed on it was changed (e.g., by changing the speed limit and/or its enforcement). An equation fitted to these data points could be properly used to estimate the mean accident rate of road segment populations that differ in mean speed and the traits associated with it; it could not be legitimately used to estimate the safety effect of changing the mean speed while keeping all other traits unchanged.

6.7.3 A Generalization

The above examples illustrate the main problem of using SPFs based on cross-section data for predicting safety effect. The core difficulty is in that when the SPF is used to predict the safety consequences of design choices or interventions one always invokes the "all-other-things-remaining-constant" assumption, whereas with SPFs based on cross-sectional data this "ceteris paribus" assumption can be seldom justified. The ceteris paribus assumption is not justified if, when predictor variable X changes, other safety-related variables that are not in the model change too.

Thus, for example, Colorado road segments found to be 0.5 miles long differ in AADT and Terrain from segments found to be 1 mile long but these traits are not now in the model equation. It follows that the curve designer may not assume that Eq. (6.1) will predict the effect of a change in tangent length as if "all other things remained constant." Similarly, roads found to have a mean speed of 55 mph differ in safety-related traits from roads found to have a mean speed of 50 mph. Therefore one may not expect that reducing the mean speed on a specific road from 55 to 50 mph by enforcement while not changing the road and its environment will increase the accident rate in accord with the fitted line in Fig. 6.8.

To justify the ceteris paribus assumption, one must be able to argue that a change in variable X now in the SPF is not associated with change in other safety-related variables that are not now in the model equation; to so argue with conviction is difficult. Even when such variables missing from the model equation do not come readily to mind, they still may exist. To illustrate, Hauer et al. (2004) developed SPFs for urban four-lane roads in which AADT, segment length, percent trucks, degree of curve, lane width, shoulder traits, driveways, and speed limit all served as predictor variables. One might think that with all these variables in the model equation an increase in speed limit would translate into an increase in the frequency of injury and fatal accidents in accord with the laws of physics and numerous empirical findings. To their surprise, the authors found that in their model, just as in Fig. 6.8, the higher the speed limit the fewer accidents were expected under otherwise identical conditions. This forced the authors to concede that *The observed relationship is unexpected if one thinks of the Speed Limit variable as accounting for the causal effect of travel speed. However, it is possible that roads where a low speed is posted were considered to be of high risk. If this judgment was correct, then the data reflect reality, only the model failed to capture some of the risk factors which were considered by those that set the speed limit.* (p. 100)

The difficulty of justifying the ceteris paribus assumption in regressions is general, besetting not only road safety, but all fields that rely on observational[28] cross-section data such as economics, sociology, and epidemiology. Naturally, therefore, there is a long standing and lively debate swirling around the causal uses of regression. What follows is a personal impression of where this debate stands.

6.7.4 The Debate

Those who design roads need to know how the choice of grade, curve radius, lane width, and of other features is likely to affect future accident frequency. Those who make recommendations about speed zoning, parking control, signal timing, or the removal or roadside obstacles need to know the safety consequences of these actions. Those who plan the transportation network need to know how the type of road and its traffic are linked to its future safety performance. Factual knowledge is the foundation of professionalism.

For many questions of this kind, the extraction of knowledge from large cross-sectional databases is the only practical approach. Nobody will contemplate randomization in road building and one cannot hope to learn from randomized experiments about, say, how a change in the radius of horizontal curves, grade, or sight distance will affect safety. Even the conduct of observational before–after studies is often difficult because, when an existing road is rebuilt, changes in radii, grades, sight distances, etc. are accompanied by a host of other changes (in road cross section, surface, clear zone width, guardrail, markings, etc.). Therefore, one

[28] Observational data: data obtained by "observing" the system of interest without subjecting it to interventions.

has to try and be clever in interpreting the kind of data we have – observational data pertaining to a cross section of units.

Unfortunately, squeezing reliable cause–effect knowledge out of observational cross-sectional data is proving to be difficult. Many have tried and accounts of their attempts can be found in the professional literature.[29] And yet, the results – while numerous – are highly variable, often mutually incompatible, and frequently contradictory (Hauer 2010). There is not the kind of reassurance that comes from the consistency and similarity of findings. Persons not steeped in the causality literature may not readily understand the difficulty. After all, if there is a causal relationship between accident occurrence and traits such as lane width or speed limit, why can one not see its reflection in the massive data sets at our disposal? The skeptical researcher will say, in answer, that while reflections of relationships can be seen in cross-section data, it is seldom clear whether they are of cause or merely of association. If cause, they are useful in predicting the consequences of road design decisions, traffic management plans, safety interventions, and planning scenarios; if associations, they are of no use for predicting the consequences of action.

To delimit the range of the debate, it is best to focus only on those purposes of regression which are controversial. Following Freedman's (1997) seminal work, Woodward[30] (2003, Sect. 7.1) lists the kinds of purposes for which regressions tend to be used:

1. To summarize or to describe a body of data.
2. To estimate the value of a dependent variable Y (here $E\{\mu\}$) from a set of independent variables $X_1, \ldots, X_n$.
3. To predict the change in the value of the dependent variable from an intervention that changes the value of X_1, etc.

Purpose "1" is uncontroversial; here the chosen model equation $\hat{E}\{\mu\} = 1.666\ X_1^{0.859}$ is only a Fisherian[31] device for the "reduction of data," a convenient tool for computing estimates which fit the 5,323 Colorado road segments. According to

[29] For a review of some historical evidence see Hauer (2005).

[30] Woodward is a philosopher with an interest in causation and explanation – a topic that has eluded consensus since Plato. In his book (Woodward 2003) emphasis is on causation in the context of manipulation and its results. To illustrate, what does the claim that coffee-drinking causes cancer say? According to Woodward, the claim means that you can affect your chances of getting cancer by changing your coffee consumption. Thus, if there is only correlation but no causation – if, for example, the statistical connection is due to some hidden factor, say stress, that causes both coffee-drinking and cancer – then you cannot increase your chances of getting cancer by drinking coffee. What causal information adds to information about correlation is information about manipulability. This kind of causality is particularly suitable for discussion of road safety because, in road safety, the question is how what we do by design and interventions is likely to affect accidents.

[31] Fisher (1922, p. 311) defines the principal task of statistics to be "the reduction of data ..."

Woodward, "*... one can use* (*the model equation*) *for this purpose even if one does not think that X_1, ..., X_n are causes of Y* (*here of* $E\{\mu\}$)" (Sect. 7.2). Using this convenient device one can say, for example, that for a 1 mile long Colorado road segment one estimates $E\{\mu\}$ to be 1.67 accidents in the 5 years 1994–1998.

Purpose "2" is to estimate the value of $E\{\mu\}$ *...given various values of X_1, ..., X_n for other bodies of data besides the data on which* (*the model equation*) *was estimated. Here again successful prediction does not seem to require that the regression equation receive a causal interpretation.* (Woodward 2003, Sect. 7.2). This purpose too is uncontroversial provided that the process generating the data serving for developing the regression equation is similar to the process where the regression equation is applied. In our case this might be to use $\hat{E}\{\mu\} = 1.667L^{0.859}$ for estimating the expected number of accidents for an *L* long segment in, say, Louisiana.[32]

The application of regression for purpose "3" is problematic. As Woodward (2003, Chap. 2) and Freedman (1997, p. 116) say: *Causal inference is different, because a change in the system is contemplated; for example, there will be an intervention. Descriptive statistics tell you about the correlations that happen to hold in the data; causal models claim to tell you what will happen to Y* (here $E\{\mu\}$) *if you change X. Indeed, regression is often used to make counterfactual inferences about the past: what would Y have been if X had been different? This use of regression to make causal inferences is the most intriguing and the most problematic. Difficulties are created by omitted variables, incorrect functional form, etc.* Elsewhere Freedman (1997, p. 59) says that: *For nearly a century, investigators in the social sciences have used regression models to deduce cause-and-effect relationships from patterns of association. ... In my view, this enterprise has not been successful. The models tend to neglect the difficulties in establishing causal relations, and the mathematical complexities tend to obscure rather than clarify the assumptions on which the analysis is based.* In our examples use of the SPF for purpose "3" would be to ask how many accidents would be saved on the tangent if curve A was replaced by curve B or how many would be caused if the speed limit was changed from 50 to 55 mph.

Why the "enterprise has not been successful" can be explained simply. Figure 6.9 shows how Hooke's Law[33] can be established: Take a spring, hang on it a weight, measure the resulting elongation; add another weight, measure again, etc. If all the data points (elongation vs. weight) form approximately a straight line, you have support for Hooke's law on this spring. The slope of the regression line fitted to these points is an estimate of the proportionality constant for this spring – the so-called spring constant. Doing the same for many other springs and if the data

---

[32] The Highway Safety Manual (AASHTO 2010) assumes that the data generating process in Colorado and Louisiana are similar except that due to local factors such as climate or procedures for accident reporting they may differ by a multiplicative "Calibration Factor" (Volume 2, p. C-4).

[33] Hooke's law is named after the seventeenth century British physicist Robert Hooke. He first stated this law in 1660 as a Latin anagram, whose solution he published in 1678 as *Ut tensio, sic vis*; literally translated as: "As the extension, so the force" or the more common meaning "The extension is proportional to the force."

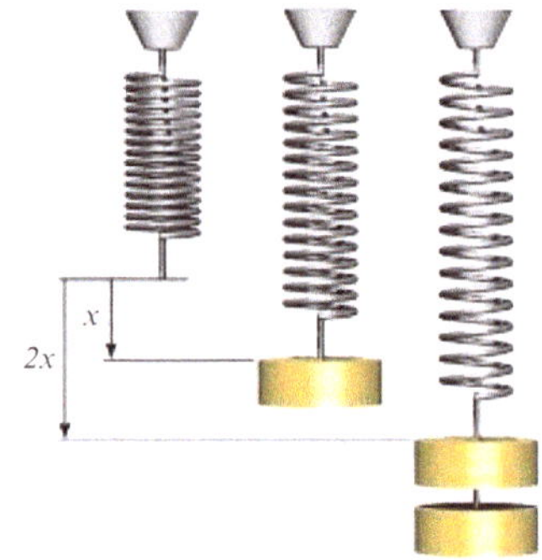

Fig. 6.9 Hooke's Law

points for each spring are close to a straight line, the law obtains support for springs in general. Once Hooke's law is shown to hold for steel springs, its validity can be similarly established for other materials and structures. With this theory in hand, bridges and airplanes can be designed.

In contrast to experimentation, imagine a roomful of springs, some with one weight attached, some with two, etc. Now the data are observational and cross-sectional. One can measure the length of each spring and note the number of weights hanging on it. If these springs and weights were known to be identical, then all springs with "n" weights would be of the same length, and Hooke's Law could again be established. That is, one could say how much longer would be such a spring made by the addition of a weight. But if the springs and the weights in the room are not certain to be identical, if springs with the same number of weights hung on them are seen to differ in length, then the task of discovering Hooke's law is more difficult, perhaps impossible. Springs seen to be long do not necessarily have many weights on them, they may have been long to begin with. Springs with many weights may be short because they are stiff or the weights light.

One could perhaps collect additional data for each spring; one could count the number of coils, measure their diameter, one could even ascertain the thickness of the wire. Using such added data, one might hope to account for the effect of visible differences between springs. That would amount to an attempt to estimate the "spring constant" from visible spring traits. This too is exceedingly difficult to do. The shear and torsion forces which elongate a spring are complex functions of the visible traits. Without reliance on theory, there is little hope that these can be successfully modeled by functions that are commonly used in regressions or by a lucky guess at what a good approximation function might be.

Then, there are also traits that are not visible. Thus, for example, two spring of identical dimensions and hung with the same weight, one made of alloy steel wire and the other from tungsten high speed steel wire will differ in length. Finally, there may be unknown-to-us reasons why one spring in the room has one weight while another spring has two or three. The "hanger of weights" may have tended to put fewer or lighted weights on the longer springs and more or heavier weights on the shorter springs. For all these reasons, the hope that Hooke's Law could be discovered from observational cross-sectional data about nonidentical springs is only a very faint one.

In road safety there are no "identical springs." This is why one should not hope that it is possible to extract trustworthy Crash Modification Factors from observational data about road segments, intersections, grade crossings, etc. Elvik (2007) discusses the difficulties which need to be surmounted for a causal interpretation of regressions in road safety to be defensible. Based on the causality criteria in epidemiology, he suggests a similar set of criteria for road safety. It so happens that these criteria are not only difficult to satisfy but also not easy to defend.[34] To illustrate his approach, Elvik (2011) chose two case studies; one where a causal interpretation is inappropriate and one where, in his opinion, "a causal interpretation is supported" (p. 261). On closer examination it turns out that even Elvik's "good regression" example cannot be causally interpreted. In the specific case which I examined (Hauer 2010) the SPFs consistently failed the basic requirements of causality.

The current state of affairs is bewildering. On one hand, *Social science (as well as psychology and epidemiology) journals are full of studies in which the presence of 'statistically significant' correlations between X and Y (perhaps when certain other variables are controlled for) is taken to establish that Xs cause or explain Ys...* (Woodward 2003, Section 4.8). The same is true for transport safety. To be convinced one only has to attend a poster session at an annual meeting of the Transportation Research Board. On the other hand the impression is that amongst eminent scholars and opinion leaders the issue has been settled a long time ago. Their conclusion: causal interpretations of multivariable regressions based on observational data are notoriously untrustworthy. Even leading advocates of the possibility to interpret observational data causally take care to distance themselves from single-equation regressions. Thus, for example, Bollen and Pearl (2013) carefully differentiate between Structural Equation Models (SEMs) and regressions. The main distinction between the two is that in an SEM the researcher must specify "what causes what" such that *each equation* (in the SEM) *is a representation of causal relationships between a set of variables, and the form of*

[34] Speaking about those coining and using the criteria for causality in epidemiology, Spirtes et al. (1993) make the following unkind comment: *Neither side understood what uncontrolled studies could and could not determine about causal relations and the effects of interventions. The statisticians pretended to an understanding of causality and correlation they did not have; the epidemiologists resorted to informal and often irrelevant criteria, appeals to plausibility, and in the worst case to ad hominem... While the statisticians didn't get the connections between causality and probability right, the... epidemiological criteria for causality were an intellectual disgrace, and the level of argument... was sometimes more worthy of literary critics than scientists.* (p. 302)

each equation conveys the assumptions that the analyst has asserted (p. 4). Furthermore, that these causal assumptions *derive from prior studies, research design, scientific judgment, or other justifying sources* and that *(t)he analysis is done under the speculation of 'what if these causal assumptions were true'*. (p. 9). Whether SEMs can or cannot be used to "*evaluate (albeit provisionally) the merits of interventional policies,*" as Bollen and Pearl claim (p. 9), is subject to vigorous and presently unsettled debate. The current differences of opinion are clearly stated in their own paper. However, there is one issue on which both sides of the debate agree, namely, that those regressions in which the model equation is chosen for its fit to the data and does not rest on causal assumptions which are justified without reliance on the same data, cannot be interpreted in cause and effect terms. In short, that . . .*no analysis in the world can derive the causal theory from non-experimental data*. (p. 29)

In parametric curve-fitting the model equation is chosen for its goodness-of-fit and parsimony of parameters. More rarely the choice is guided by the quality of the estimates of $E\{\mu\}$. If so, at this time, the model equation is not the embodiment of some commonly accepted theory or preexisting body of consistent empirical evidence and therefore does not amount to a "causal law," "understanding," or an "explanation."[35] Nor can one be convinced that if information about other safety-related variables was available and added to the model equation, the parameter estimates would not substantially change. As Davis (2014) notes, the extent to which an observed association is accepted as causal depends on the extent to which alternative explanations for the association can be rejected. Since alternative explanations can only seldom be ruled out, it is perilous to use SPFs for predicting the consequences of interventions or design changes. Why then, without batting an eye, do so many researchers present their parametric regressions models as a guide to action or source of Crash Modification Factors?

Hope springs eternal. Because cause–effect relationships are surely embedded in the plentiful data sets available, and because searching for these relationships is in many cases the only practical way to the knowledge needed, it is reasonable to persevere; to seek paths through the thicket, and to expect that the clearings of knowledge will eventually be visible. The dream is that, eventually, by improvements in methods of modeling and in the quality of data, research will begin to yield compatible and consistent results. If and when they do, this jaundiced view of the use of SPFs for assessing the safety effect of design decisions and interventions will have to be revised.

Having discussed what curve-fitting can and may not provide, in the next chapter we return to the task of SPF development. Consistent with the focus and perspective

[35] An explanation is a set of statements constructed to describe a set of facts which clarifies the causes, context, and consequences of those facts. Explanation and cause seem to be inseparable (see, e.g., Psillos 2002) Unfortunately statisticians speak about "explained variance" in a manner divorced from causation, forgetting their own dictum that "association is not causation."

chosen Sect. 1.5 the purpose of SPF development is to provide good estimates of $E\{\mu\}$ and $\sigma\{\mu\}$ in support of the applications described in Sect. 1.3.

6.8 Summary

The process of SPF development entails the choosing of the traits (variables) from which the model equation is to be made, of determining the form of the function representing the variables in the model equation, of estimating the value of the parameters which best fit the available data, of examining the "goodness" of the fit and of the $E\{\mu\}$ estimates, and of searching for ways to improve them. This process is one of trial and error, of groping and backtracking, and favors the gradual buildup of the model equation. In this "first steps" chapter "Segment Length" was chosen to be the first variable introduced into the model equation and the function $\beta_0(\text{Segment Length})^{\beta_1}$ was used to represent its influence on $E\{\mu\}$. The Solver found those values of β_0 and β_1 for which the sum of (weighted) squared differences was the smallest.

A method for the estimation of $\sigma\{\mu\}$ was presented. Once again nonparametric regression proved useful in revealing the regularity in what seemed to be a patternless cloud of data points. In the Colorado data set the estimate of $V\{\mu\}$ was proportional to the square of the estimate of $E\{\mu\}$. Thus, at the end of Sect. 6.5, the first complete SPF could be written.

The C-F spreadsheet proved to be a flexible and user-friendly environment for SPF development. It was simple to code in formulae for fitted values or elements of the objective function and, when modifications were needed (such as when weights had to be computed), these were easily accommodated.

The "first steps" nature of the SPF, because of its transparency, was well suited for raising two general questions. One was about the inaccuracy of parameter estimates, the other about whether an SPF can be used to predict the safety effect of design changes and interventions.

The variability of parameter estimates is made of two parts: the "modeling" and the "statistical" inaccuracies. Modeling inaccuracy reflects the variability of parameter estimates which comes from the choices which the modeler makes; it depends on which variables are represented in the model equation, by what function they are represented, and what is minimized or maximized when parameters are estimated. The statistical part of parameter inaccuracy reflects the variability of the data; it may come from uncertainty about variable values (e.g., of AADT estimates) or from the randomness inherent in accident counts.

The statistical inaccuracy of a parameter estimate is quite easy to quantify by Monte Carlo simulation. In contrast, that part of that variability which is due to omitted variables and other modeling choices is not usually known and cannot be reported on. Because "statistical accuracy" is only one part of the picture, it is an underestimate and reporting it without qualification misleads.

The question of the causal uses of SPFs is discussed in Sect. 6.7. Because regression equations look like mathematical equations that represent functional dependence, the difference between the two had to be clarified. A functional dependence does not imply causation; an equation representing causation cannot be manipulated as if it was a function. Two examples are furnished to illustrate the point. The first shows why our first SPF may not be used in road design, the second shows why a curve fitted to data may not be given causal meaning.

When an SPF is used to predict the safety consequences of some design choice or of an intervention one always assumes that "all other thing remain constant." With SPFs based on cross-sectional data this ceteris paribus assumption can be seldom justified. To justify it, one should be able to argue that a change in a variable of the SPF is not associated with change in other safety-related variables that are not now in the model equation.

Why it is so difficult to reach causal conclusions using cross-sectional data has been illustrated with reference to Hooke's law. Opinion leaders in other fields have long recognized the difficulties that face the causal interpretation of cross-sectional data by single-equation regressions and reached their verdict. It is therefore surprising that the same is not more widely acknowledged in the road safety literature.

As noted at the beginning of this summary, there are many choices which the modeler has to make. Whether a modeler's choice made for a better model is often judged by whether there is an improvement in the fit of the model to the data. How to judge the quality of a fit will be discussed next.

References

The American Association of State Highway and Transportation Officials (AASHTO) (2010) Highway safety manual, 1st edn. Washington, DC, AASHTO

Aitken AC (1935) On least squares and linear combinations of observations. In: Proceedings of the Royal Society of Edinburgh, 55:42–48.

Bollen KA, Pearl J (2013) Eight myths about causality and structural equation models. In: Morgan S (ed) Handbook of causal analysis for social research. Springer, New York

Davis GA (2014) Crash reconstruction and crash modification factors. Accid Anal Prev 62:294–302

Elvik R (2007) Operational criteria of causality for observational road safety evaluation studies. Transp Res Rec 2019:74–81

Elvik R (2011) Assessing causality in multivariate accident models. Accid Anal Prev 43:253–264

Fisher RA (1922) On the mathematical foundations of theoretical statistics. Philos Transact A Math Phys Eng Sci 222:309–368

Freedman D (1997) From association to causation via regression. Adv Appl Math 18:59–110

Garber NJ, Gadirau R (1988) Speed variance and its influence on accidents. AAA Foundation for Traffic Safety, Washington, DC

Hauer E (2005) Cause and effect in observational cross-section studies on road safety. https://www.researchgate.net/profile/Ezra_Hauer/ publications

Hauer E (2010) Cause, effect, and regression in road safety: a case study. Accid Anal Prev 42:1128–1135

Hauer E, Council FM, Mohammedshah Y (2004) Safety models for urban four-lane undivided road segments. Transp Res Rec 1897:96–105

JCGM (Joint Committee for Guides in Metrology) (2008) International vocabulary of metrology—basic and general concepts and associated terms. 200:2008

Mensah A, Hauer E (1998) Two issues of averaging in multivariate modelling. Transp Res Rec 1635:37–43

Psillos S (2002) Causation and explanation. McGill-Queen's University Press, Montreal

Simon HA (1953) Causal order and identifiability. In: Hood W, Koopmans T (eds) Studies in econometric method, Cowles commission monograph 14. Yale University Press, New Haven, pp 49–74

Spirtes P, Glymour C, Scheines R (1993) Causation, prediction and search, lecture notes in statistics, vol 81. Springer, New York

Straume M, Johnson ML (2010) Monte Carlo method for determining complete confidence probability distributions of estimated model parameters. In: Johnson M (ed) Essential numerical computer methods. Academic, New York

Woodward J (2003) Making things happen: a theory of causal explanation. Oxford University Press, Oxford

Which Fit Is Fitter

7

Abstract

In this chapter discussion is about how well a model fits the data. There are many single-number goodness-of-fit measures but they all describe only the overall fit. For SPF modeling, this is insufficient. For the SPF to produce useful estimates, they must be good for all values of every variable. An alternative tool to describe goodness of fit is suggested and its uses explained. The CURE plot shows at a glance how good a fit is and what the remaining concerns are. Long up or down runs indicate regions of bias which demand model improvement either by the addition of new traits or by a change of functional form; large vertical drops in the CURE plot invite the examination of outliers. The CURE plot is useful in determining whether a fit is acceptable and in judging which of two fits is better.

7.1 Goodness of Fit

The modeler aims to develop an SPF which fits the data well.[1] To do so one has to be clear about what makes for a good fit. The fit of a model is judged by its residuals—the differences between the number of recorded accidents and the

[1] While this sentence seems unobjectionable, a comment is in order. If the aim is indeed that of getting a good fit then why, when estimating parameters, the objective is not that of maximizing fit? The answer is that the usual focus in regression modeling is that of getting good parameter estimates, not good estimates of $E\{\mu\}$. This dichotomy and choice of perspective was discussed in Sect. 1.5 and its implication for the choice of objective function will be further explored in Chap. 8. Here it is sufficient to note that in traditional regression modeling there is a tension between estimating parameters with one purpose in mind and judging the quality of the model with a different yardstick in hand.

E. Hauer, *The Art of Regression Modeling in Road Safety*,
DOI 10.1007/978-3-319-12529-9_7

number fitted by the model equation.[2] A model is thought to fit well if the residuals are "tightly packed" around zero.

There are many ways in which this tightness of packing is measured. Commonly used are single-number measures of goodness of fit such as the χ^2, the coefficient of determination (R^2), the scaled deviance (G^2), and similar statistics.[3]

Single-number measures describe an "overall goodness of fit." For the purposes for which SPFs are developed, the "overall goodness of fit" is of little interest. The SPF is to produce good estimates of $E\{\mu\}$ and $\sigma\{\mu\}$ for all variable values of interest.[4] An SPF that overestimates $E\{\mu\}$ in one range of a variable and underestimates it elsewhere may fit well overall but is "biased everywhere" and, as such, of little practical use. To illustrate, the hypothetical plot of residuals in Fig. 7.1 indicates that for variable values between about 30 and 70 the fitted value is too small and the opposite is true elsewhere; this model may be satisfactory overall but is biased everywhere. In SPF development the main figure of merit is the absence of bias for all variable values for which the model is likely to be used.

The standard procedure for examining residuals in detail is to plot the residuals on the vertical axis against trait levels (variable values) on the horizontal axis.[5] Figure 7.2 is such a plot of the 5,323 residuals for the weighted least-squares fit in Fig. 6.4. The insert shows details near the origin.[6] As is evident, the standard plot of residuals is not informative. Without some smoothing, it is difficult to see whether the fit is unbiased everywhere and whether it is well packed around the horizontal axis. A different approach is needed.

---

[2] There is in use an annoying multiplicity of terms which all seem to have the same meaning. In curve-fitting one tends to use "difference" and "deviation"; both stand for the D in acronyms such as SD and SSD. In examining the goodness of a fit the word "residual" is usually used to mean the same. I will shun the word "deviation" and, reluctantly, use "difference" for the D in acronyms and "residual" to mean the same when speaking about the quality of a fit. There is yet another related and easily confused term in common use: "error." While "difference" and "residual" always refer to "observed-fitted," "error" refers to "observed-expected."

[3] Other commonly used goodness-of-fit measures are the Root Mean Square Error of Approximations, Statistical Deviance, the AIC (Akaike Information Criterion), the BIC (Bayesian Information Criterion), etc. See, e.g., Miaou (1996, Miaou et al. 1996), Schermelleh-Engel et al. (2003), and Hooper and Mullen (2008).

[4] The question of what SPFs are for was discussed in Chap. 1. More specifically the differences between the "applications" and the "research" perspectives were described in Sect. 1.5. For this book the "applications" perspective was chosen and, as a consequence, emphasis is on how well $E\{\mu\}$ and $\sigma\{\mu\}$ are estimated, not on the accuracy of the parameter estimates. This shift in emphasis is reflected in modeling. Here it leads to the abandonment of single-number measures of goodness of fit.

[5] See, e.g., Draper and Smith (1981, p. 148).

[6] The striation is due to the definition residual $\equiv$ observed-fitted in which the observed is always an integer.

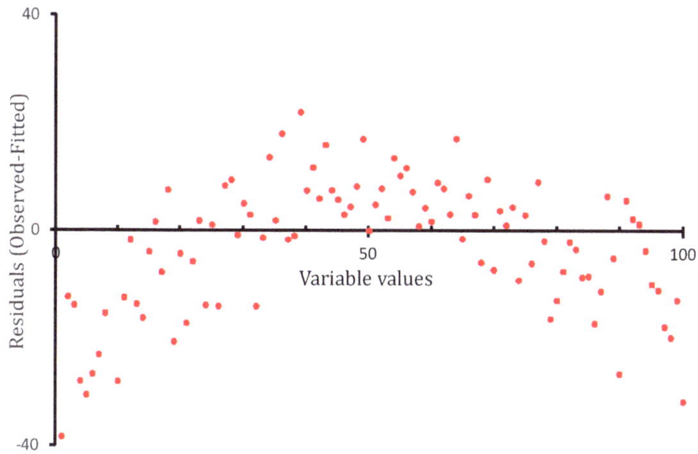

Fig. 7.1 Plot of residuals

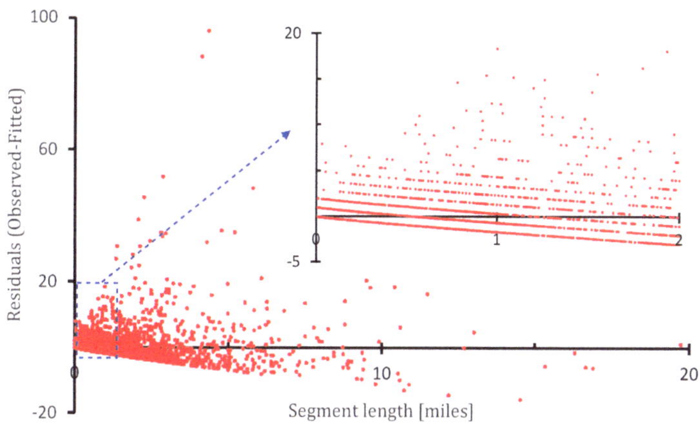

Fig. 7.2 Residuals for Fig. 6.4

7.2 The CURE Plot[7]

Plotting cumulative residuals is better. As shown in Fig. 7.3 before the computation of cumulative of residuals can commence the data (copied from the C-F spreadsheet in Fig. 6.4) have to be sorted in the ascending order of the variable of interest. Here the data are sorted by Segment Length ("Miles" in column B). When the residuals

[7] The CURE method is described in Hauer and Bamfo (1997). Because the reference is not easily accessible the derivation of the main equation (7.1) is in Appendix I.

Data from figure 6-4

	A	B	C	D
1	**CDOT, Two-lane, Rural**			
2			Observed	wSSD
3	Segment	Miles	I&F Sum 94-98	Fitted Value
4	1	1.50	0	2.360
5	2	2.90	0	4.159
6	3	1.49	0	2.347
5325	5322	0.14	0	0.308
5326	5323	0.48	3	0.887

Sorted by 'Miles'

	A	B	C	D
1	**CDOT, Two-lane, Rural**			
2			Observed	wSSD
3	Segment	Miles	I&F Sum 94-98	Fitted Value
4	107	0.01	0	0.032
5	331	0.01	0	0.032
6	404	0.01	0	0.032
5325	5115	16.9	13	18.915
5326	4160	19.74	23	21.617

With Cumulated Residuals

	A	B	C	D	E	F	G	
1	**CDOT, Two-lane, Rural**							
2			Observed	wSSD	Σfitted value	Residuals	ΣResiduals	
3	Segment	Miles	I&F Sum 94-98	Fitted Value		Obs.- Fitted		
4	107	0.01	0	0.032	0.032	-0.03	-0.03	
5	331	0.01	0	0.032	0.064	-0.03	-0.06	
6	404	0.01	0	0.032	0.096	-0.03	-0.10	
2614	3897	0.83	13	1.42	1579.47	11.58	293.53	
2615	4227	0.83	2	1.42	1580.89	0.58	294.11	A
2616	4424	0.83	1	1.42	1582.31	-0.42	293.69	
5325	5115	16.9	13	18.915	9207.361	-5.92	-1.36	
5326	4160	19.74	23	21.617	9228.977	1.38	0.02	

Fig. 7.3 The CURE spreadsheet: computing cumulated residuals.[8]

are so cumulated and the ordinates in column G are plotted against the abscissae in column B Fig. 7.4 obtains.[9] Plots of this kind will be called CURE (CUmulative REsiduals) plots.

The benefits of residual cumulation are several. First, the patternless chaos of Fig. 7.2 is replaced by instant clarity in Fig. 7.4. The rising pieces of the jagged curve (Origin to A, B to C and E to F) correspond to segment lengths where the observed accident counts tend to be larger than what the model predicts; here the model equation underestimates. Conversely, segment lengths over which the jagged-curve decreases (here precipitously) is where the model overestimates. This model is approximately unbiased only between C and D; it is biased everywhere else. Obviously the fitted model equation 1.666 (Segment Length)$^{0.859}$ is not a good way to estimate $E\{\mu\}$. Whether the model can be improved by using a different objective function, by choosing a more suitable functional form, by adding variables, and by removing outliers, remains to be seen.

---

[8] To download this spreadsheet, go to http://extras.springer.com/ and enter the ISBN of this book. Look in the "Spreadsheets" folder for "Chap. 7. CURE computations .xls or xlsx"

[9] To show the relevant detail clearly, segment length was truncated at 3 miles leaving off the longer 5 % of segments.

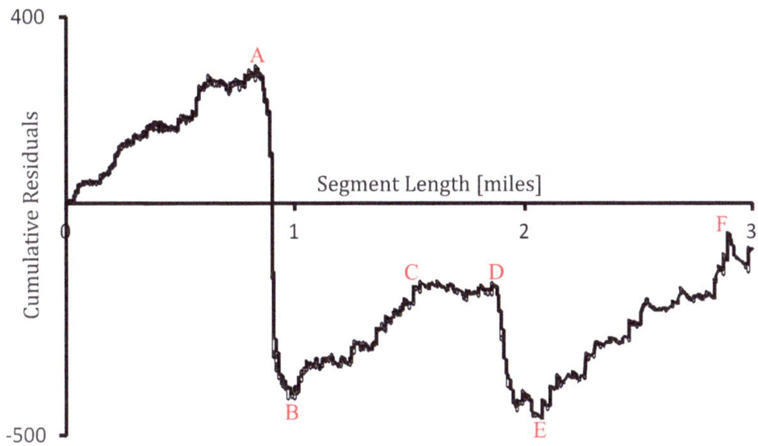

Fig. 7.4 CURE plot for Segment Length

Second, it is now easy to say what a good CURE plot should look like. It should not have vertical drops because these are indicative of inordinately large residuals—possible outliers. It should not have long increasing or decreasing runs because these correspond to regions of consistent over and underestimation. It should meander around the horizontal axis in a manner consistent with a "symmetric random walk."[10] The CURE plot in Fig. 7.4 does not come close to this desideratum.

Third, the cost of parametric curve-fitting is now manifest. As noted earlier, the assumption that underneath the cloud of data points there is some simple and smooth mathematical function is an act of faith. Without such an assumption there never can be enough data to estimate the $E\{\mu\}$ of a population when it is defined by more than two or three safety-related traits. What is that smooth function is never known. It is the modeler who selects the function to be fitted. As the CURE plot in Fig. 7.4 clearly shows, the Colorado data cannot be adequately represented by the $\beta_0(segment\ length)^{\beta_1}$ function. Imposing this function on the data produces estimates of $E\{\mu\}$ that are biased at almost all values of Segment Length. Should a modeler choose such an unsuitable function, either by convenience or by rote, and when the results are later published and used, society pays the cost. The cost is

---

[10] In a one-dimensional random walk one follows the evolution of the sum of independent random variables. Let $R_1, R_2, \ldots, R_N$ be a sequence of N independent random variables. The sequence of points with coordinates $\left(i, \sum_1^i R_i\right)$ can be visualized as a random walk. It is what the movement over time of a stock price would look if the probability of a price increase was the same as the probability of an equal price drop, and if stock price changes over time were statistically independent. The random walk of interest here is one where the Rs correspond to residuals. When the residual is positive, the sum moves up, with a negative residual it moves down. Inasmuch as for every i the sum of all residuals is expected to be 0, this kind of random walk oscillates around the horizontal axis.

that of putting into the practitioner's hands a tool that produces biased estimates. Biased estimates may lead to bad decisions. Bad decisions have real costs; costs that are attributable to the use of an unsuitable model equations. The need to use a model equation is the hallmark of parametric curve-fitting.

Fourth, the CURE plot may provide clues about what needs improving and what the agents of improvement might be. Where the fitted values are too low (Origin to A, B to C and E to F) they must be raised. A function capable of taking this shape should replace the current model equation. Where there is a precipitous drop outliers may be hidden. A search for outliers in these spots might help.

The CURE plot in Fig. 7.4 has Segment Length on the horizontal axis and shows how well or how poorly the SPF would predict for various values of this variable. As other variables will be added to the model equation, each will have its own CURE plot. These variable-based CURE plots will be used to examine the goodness of fit for each variable and to examine ways in which the fit for that variable could be improved. Having another variable in the model equation becomes manifest in the CURE spreadsheet of Fig. 7.3 as another data column placed before the observed accident counts (there in column C). To produce a CURE plot for a new variable, all that needs to be done is to sort the columns up to and including the fitted values the ascending order of that variable.

After the diagnostic clues of all variable-based CURE plots are exploited, a fitted value-based CURE plot will be prepared. Here the sorting is by the fitted value column and the plot shows how well or poorly the SPF predicts, not for a specific variable but overall, as a function of number of accidents expected on a unit.

7.3 The Bias-in-Fit

The absence of bias was said to be the main figure of merit for an SPF intended for use in applications. A horizontal stretch of the CURE plot corresponds to a region of the variable where the estimates are unbiased. Conversely, in regions where the CURE plot consistently drifts up or down the estimates are biased. The systematic discrepancy between the observed and the fitted values will be called the bias-in-fit.[11] To describe and quantify the bias-in-fit the CURE plot in Fig. 7.4 is approximated by a sequence of straight lines such as the polygon 0-A-B-C-D-E-F in Fig. 7.5.

Using the data from the CURE spreadsheet in Fig. 7.3, at point A (Segment Length = 0.83 miles, cumulative residual = +294) the accumulated number of fitted accidents is 1,581. Thus, for segments in the 0–0.83 miles range the fitted values underestimates the observed values on the average by about $100 \times (294/1{,}581) = 18.6\ \%$. Similar computations for the other ranges of Segment Length are in Fig. 7.6. In most ranges the % Bias/Accident is unacceptably large.

[11] Another kind of bias, the "bias-in-use," will be introduced and discussed in Chap. 10. Bias-in-use occurs when the information about variables which is available to the user of the SPF differs from the variables in the SPF.

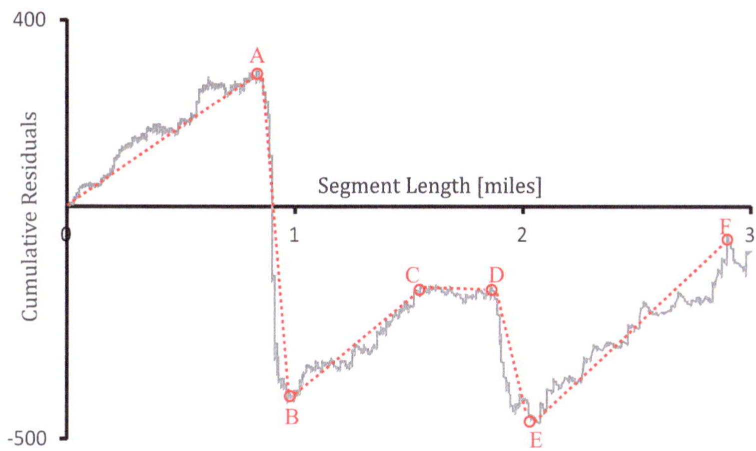

Fig. 7.5 Replacing the CURE plot by a polygon

	Data			Computations			
	Miles	Σ Fitted	Σ Residuals	Bias in range	Fitted in range	% Bias per Accident	Cumulated Abs(Bias)
Origin	0.00	0	0				
				294	1581	18.6	294
A	0.83	1581	294				
				-714	1598	-44.7	1008
B	1.00	3178	-419				
				238	1094	21.8	1246
C	1.53	4272	-181				
				9	443	2.0	1255
D	1.86	4716	-173				
				-293	781	-37.5	1547
E	2.08	5496	-465				
				401	928	43.2	1948
F	2.89	6424	-64				
				64	2805	2.3	2013
End	19.74	9229	0				TAB

Fig. 7.6 Bias-in-fit analysis

An interesting figure of merit is the accumulated absolute bias in the rightmost column of Fig. 7.6. Inasmuch as the absence of bias is a sensible yardstick for measuring model performance, the Total Accumulated (absolute) Bias can be used for model fitting. In Fig. 7.6 the total accumulated bias (TAB) is 2013.

The next question is how to distinguish between a satisfactory CURE plot, one that is close to a symmetric random walk, and an unsatisfactory one that is indicative of the presence of bias-in-fit. Before tackling this question a minor obstruction needs removing.

7.4 Leveling the Playing Field

In Fig. 6.4 the sum of observed accidents was 9,229 and the sum of fitted accidents happened to be nearly the same (9,228.98). This degree of agreement is rare. Thus, for example, in Fig. 6.3 where a different objective function was used, the sum of fitted accidents was 9,309.7 differing substantially from the 9,229 observed. Discrepancies between the sum of fitted and observed accidents occur always when models are fitted by minimizing or maximizing some objective function. This discrepancy is a bother when one wishes to compare the goodness of fit of two models. The same discrepancy is also an obstacle to obtaining a simple expression for the limit which a genuine random walk is unlikely to exceed. This bother and obstacle can be eliminated by adding a constraint to the Solver.[12] How to add such a constraint is shown in Fig. 7.7.

Now the road is clear to determining the acceptability of a CURE plot and to deciding which two plots fit is better.

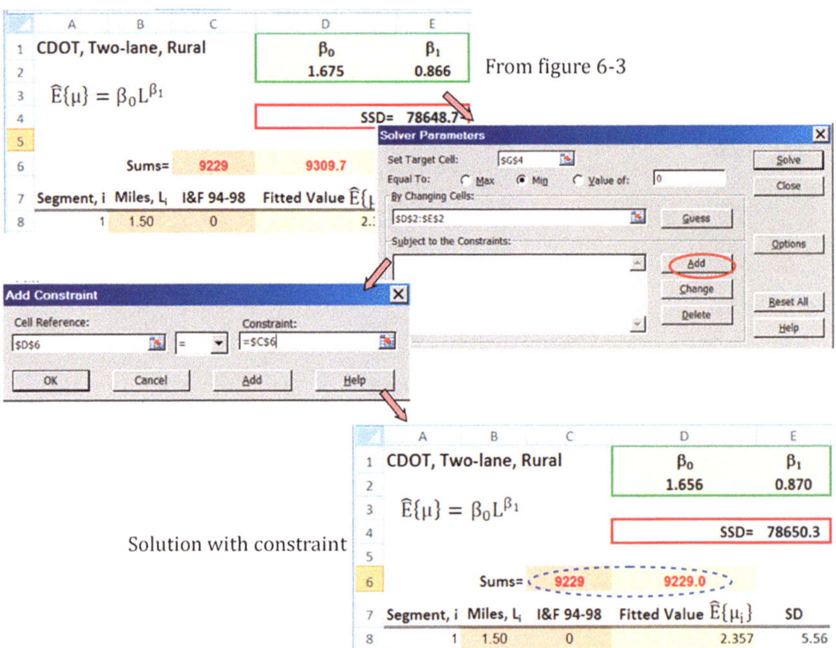

Fig. 7.7 How to add a constraint

[12] The addition of this constraint will cause a slight increase of the objective function when minimized and a small decrease when maximized. Thus, for example, in Fig. 7.7 the SSD increased from 78,648.7 to 78,650.3.

7.5 When Is a CURE Plot Good Enough?

Where a CURE plot consistently drifts upward or downward there is bias-in-fit. However, even in the absence of such bias, random walks will have "runs," i.e., stretches in which several consecutive residuals tend to be positive (an up-run) or negative (a down-run). A tool is needed to distinguish between runs that are consistent with the play of chance in an unbiased model and those that indicate the presence of bias-in-fit.

To help with the distinction, one can compute the limits beyond which a random walk of an unbiased CURE plot should only rarely go.[13] How to do so is shown on the "CURE spreadsheet" in Fig. 7.8. The top panel comes from Fig. 7.3 and is the source of the auxiliary computation (columns H to K) in the middle panel. The estimates of the standard deviation for an unbiased random walk constrained to end on the horizontal axis, the $\pm\hat{\sigma}_s^{'}(i)$, is in column L of the bottom panel. The $\perp 2\hat{\sigma}_s^{'}(i)$ limits are in columns M and N. The corresponding CURE plot is in Fig. 7.9.

The reasoning behind the computation is straightforward. Let $i = 1, 2, \ldots, n$ be the index of the sorted rows. The ordinate of the CURE plot at i is the sum of i independent residuals. As the residuals are random variables, so is their sum $S(i)$. Being a random variable $S(i)$ has a mean and a standard deviation. If the model produces estimates of $E\{\mu\}$ that are unbiased everywhere, the mean of $S(i)$ is 0 for all i. However, not so is the standard deviation of $S(i)$ denoted by $\sigma_S(i)$. This value can be estimated by the square root of the sum of squared residuals as was done in column K. Let $\sigma_S^{'}(i)$ denote the standard deviation of $S(i)$ for those random walks that end on the horizontal axis (i.e., that $S(n) = 0$.) In Appendix I it is shown that

$$\pm\sigma_s^{'}(i) = \pm\sigma_S(i)\sqrt{\left(1 - \frac{\sigma_S^2(i)}{\sigma_S^2(n)}\right)} \tag{7.1}$$

To illustrate consider row 2500 in Fig. 7.8. Here the index $i = 2{,}497$ and the sum of squared residuals is 3,898. In the last row $i = n = 5323$ and the sum of squared residuals is 78,656. By (7.1), $\pm\hat{\sigma}_s^{'}(i) = \sqrt{3,898}\sqrt{1 - 3,898/78,656} = \pm 60.9$. If most symmetric random walks are within two standard deviations from the horizontal axis, we expect the sum of residuals at this point to be within ±122. As it happens the sum of residuals at this point is 256, far outside the limits, a clear indication of pervasive bias-in-fit in this model.

[13] Here only the procedure for computing the limits will be described. The derivation of the expression by which the limits are computed is in Appendix I. To download this spreadsheet, go to http://extras.springer.com/ and enter the ISBN of this book. Look in the "Spreadsheets" folder for "Chap. 7. CURE computations .xls or xlsx".

	A	B	C	D	E	F	G
2			Observed	wSSD	Σfitted value	Residuals	Σ Residuals
3	Segment	Miles	I&F Sum 94-98	Fitted Value		Obs.- Fitted	S(i)
4	107	0.01	0	0.03	0.03	-0.03	-0.03
5	331	0.01	0	0.03	0.06	-0.03	-0.06
2499	4790	0.77	0	1.33	1420.77	-1.33	252.23
2500	5247	0.77	5	1.33	1422.10	3.67	255.90
2501	104	0.78	0	1.35	1423.45	-1.35	254.55
5325	5115	16					-1.36
5326	4160	19.					0.02

	H	I	J	K
1		Auxiliary Computations		
2	Index	Squared	Cumulated	$\hat{\sigma}_S(i) = \sqrt{\hat{\sigma}_S^2(i)}$
3	i	Residual	$\hat{\sigma}_S^2(i)$	
4	1	0.0010	0.0010	0.03
5	2	0.0010	0.0020	0.05
2499	2496	1.7712	3884.2322	62.32
2500	2497	13.4626	3897.6949	62.43
2501	2498	1.8109	3899.5057	62.45
5325	5322	34.9927	78654.1330	280.45
5326	n=5323	1.9138	78656.0468	280.46

	L	M	N
1			
2	$\hat{\sigma}'_S(i) = \pm\hat{\sigma}_S(i)\sqrt{1 - \frac{\hat{\sigma}_S^2(i)}{\hat{\sigma}_S^2(n)}}$	Upper	Lower
3		Limit	Limit
4	0.03	**0.06**	**-0.06**
5	0.05	**0.09**	**-0.09**
2499	60.77	**121.5**	**-121.5**
2500	60.87	**121.7**	**-121.7**
2501	60.88	**121.8**	**-121.8**
5325	1.38	**2.77**	**-2.77**
5326	0.00	**0.00**	**0.00**

Fig. 7.8 The CURE spreadsheet

Inasmuch as the CURE plot is a sum of many independent random variables, it is approximately normally distributed.[14] For a normal distribution, about 95 % of the probability mass is within two standard deviations from the mean. Thus, the CURE plot for an "everywhere unbiased" SPF should only rarely go beyond the $2\sigma'$ limits;

[14] In probability theory, the central limit theorem states that, given certain conditions, the sum of a sufficiently large number of independent random variables, each with finite mean and variance, will be approximately normally distributed.

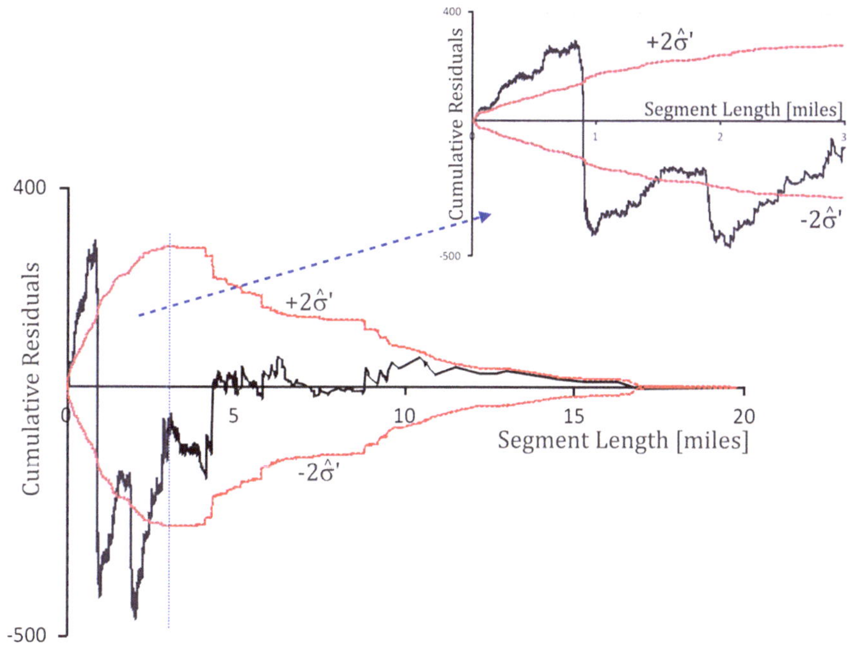

Fig. 7.9 CURE plots with limits

if it does, improvements should be sought. Using the same reasoning, about 40 % of the Normal probability mass is within half a standard deviation from the mean. It follows that the CURE plot for many unbiased models can be expected to exceed the $\pm 0.5\sigma'$ limits. Models the CURE plot of which does not go beyond the $0.5\sigma'$ limits are close to being unbiased and attempts to further "improve" such models court the danger of "overfitting." With this guidance one can decide whether a model requires improvement and whether a model is good enough to be left alone. For in-between models, those for which the CURE plot punctures the $0.5\sigma'$ boundary but not the $2\sigma'$ boundary, improvement may be attempted.

7.6 Comparing CURE Plots

In Sect. 6.5 two sets of parameter estimates for the model equation $\beta_0 X_1^{\beta_1}$ were obtained. When the weight was computed using the first fitted value the estimates of β_1 and β_2 were 1.666 and 0.859; when the weight was allowed to converge to the current fitted value, the estimates were 1.766 and 0.737. The question is which set of parameter estimates makes for a better fit. The corresponding CURE plots are in Fig. 7.10.

Two observations follow. First, that one can tell the difference between a better and a worse fit. Such judgments must be based on an examination of the bias-in-fit,

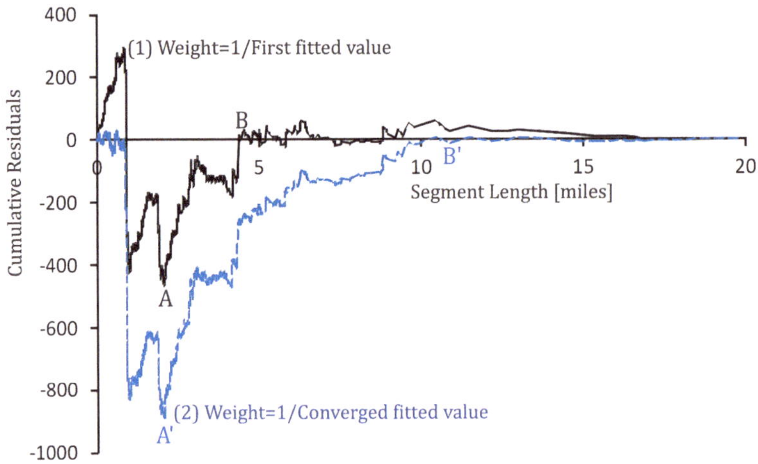

Fig. 7.10 Comparison of two CURE plots

not on an impression of proximity to the horizontal axis. In Fig. 7.10 the bias-in-fit between A and B is smaller than between A' and B'[15] which is why fit (1) is better than fit (2). Second, the two plots are nearly parallel. That means that no matter what β_0 and β_1 are chosen, the $\beta_0 X_1^{\beta_1}$ model equation will have similar bias-in-fit problems. Choosing a different objective function while keeping the same model equation will not help. What may help is using a different model equation, adding variables to it, and removing outliers.

7.7 Summary

The outlines of the modeling process are gradually emerging. In Chap. 6 a bare-bones model equation was introduced and parameter estimation on a C-F spreadsheet was demonstrated. In this chapter discussion was about how well this embryonic SPF fits the data. In the coming chapters the model will be elaborated on and improved.

The fit of a model is judged by its residuals. In the literature one finds an abundance of "goodness-of-fit" measures. The problem is that they all describe only how the model fits overall and do so by a single number. For applications oriented SPF modeling, this is insufficient. To be of practical use, the estimates which the SPF produces have to be nearly unbiased for all variable values.

To examine the fit of a model in the requisite detail, it is commonly recommended to plot residuals against variables of interest. Unfortunately, for

[15] Because at its right end the CURE plot must land on the horizontal axis the underestimation of the fitted values in the AB (or A′B′) range is always accompanied by a compensatory overestimation elsewhere.

SPFs this plot is difficult to interpret. A better alternative is to use cumulative residuals, the CURE plot.

How to prepare CURE plots was explained. The CURE plot shows at a glance whether there are concerns and what they are. Long up or down runs indicate regions of bias-in-fit. Vertical drops need to be examined for the presence of outliers.

A method for estimating the magnitude of bias-in-fit was provided. The presence of significant bias-in-fit is the hallmark of an ill-fitting model. Ill-fitting models need improving by the addition of new variables, changes of functional form, and removal of outliers.

Whether a model is free of significant bias can be judged with reference to the limits which an unbiased random walk of residuals is unlikely to exceed. A way to compute these limits was provided. The CURE plot was also shown to be useful in judging which of two fits is better.

Here is where we now stand. The fit of the bare-bones model produced so far was found to be unsatisfactory. The hope is that the addition of variables (AADT, Terrain), the use of more appropriate functions in the model equation, and the removal of outliers will improve fit. It is also possible that the fit is poor partly because it was obtained by minimizing the weighted sum of squared differences. Even though this is a popular minimand, especially in econometric modeling, there are grounds to think[16] that other objective functions are more suitable for SPF modeling. Perhaps if a different minimand or maximand was used, the fit would be better. The question of what should be optimized will be pursued next.

References

Draper N, Smith H (1981) Applied regression analysis, 2nd edn. Wiley, New York

Hauer E, Bamfo J (1997) Two tools for finding what function links the dependent variable to the explanatory variables. Proceedings of the ICTCT 97 conference, Lund, Sweden, pp 1–19

Hooper DC, Mullen MR (2008) Structural equation modelling: guidelines for determining model fit. Electron J Bus Res Meth 6(1):53–60, Available online at www.ejbrm.com

Miaou S-P (1996) Measuring the goodness-of-fit of accident prediction models. FHWA-RD-96-040. Federal Highway Administration, Office of Safety and Traffic Operations, Washington DC

Miaou S-P, Lu A, Lum HS (1996) Pitfalls of using R^2 to evaluate goodness of fit of accident prediction models. Transp Res Rec 1542:6–13

Schermelleh-Engel K, Moosbrugger H, Müller H (2003) Evaluating the fit of structural equation models: tests of significance and descriptive goodness-of-fit measures. Meth Psychol Res 8 (2):23–74

[16] One such reason is that the distribution of crash counts is skewed while least-squares curve-fitting is suitable for symmetrical distributions.

What to Optimize?

8

Abstract
Parameters estimated by weighted least squares produced bad fits. The question is whether the fit can be improved by maximizing likelihood or using some alternative objective function. The concepts of likelihood, likelihood function, and maximum likelihood estimation will be explained and illustrated. In preparation for model development some commonly used likelihood functions will be given and implemented on a C-F spreadsheet. When the purpose of SPF development is to support practical road safety management, quality of fit rather than maximization of likelihood may be the preferred objective. The use of such alternative objective functions will prove both simple and attractive.

8.1 Introduction

Optimization is the Archimedean lever used to lift parameters out of their obscurity. In Chap. 6 parameters were estimated by minimizing the weighted sum of squared differences. The quality of the fit was examined in Chap. 7 and was found wanting. Perhaps the fit was bad because the chosen model equation cannot take on the form that the Colorado data call for; perhaps it was bad because important traits were not yet accounted for in the model equation, perhaps it was bad because of outliers. Whether the fit will improve when a different model equation will be used, variables added and outliers removed will be examined later. In this chapter the question is whether the fit can be improved by using a more suitable objective function. What objective functions should be tried? How can these be accommodated in a C-F spreadsheet? Do they yield better fits? These are the issues to be aired.

E. Hauer, *The Art of Regression Modeling in Road Safety*,
DOI 10.1007/978-3-319-12529-9_8

Parameters are usually estimated by either minimizing the sum of (weighted) squared differences or by maximizing likelihood.[1] Both approaches have deep roots in the history of statistical thought; both have strengths and weaknesses.[2] The weakness of the least squares approach comes from the diversity μ's amongst the units that serve as data. Units that have different μ's differ in the variances of the observed accident counts. This is a problem because for estimation by least squares one assumes that the variances are all the same. To remove this obstacle one may resort to weighted least squares estimation but the determination of the weights can be problematic.[3]

The weakness (or perhaps the strength?) of the likelihood approach is in the need to make specific, detailed, and explicit assumptions about the probability distribution that generated the accident counts. As one does not know how well these assumptions approximate reality, it is difficult to distinguish between a good result and one that should not be trusted. Still, estimating parameters by maximizing likelihood is currently the dominant approach in SPF modeling. This is why pains

---

[1] In the mainstream statistical thought the least squares and maximum likelihood rationales are motivated by the desire to get parameter estimates with desirable properties such as being unbiased, having minimal variance etc. Parameters are thought to be the reflection of some underlying law-like regularity present in nature or in human affairs which the researcher aims to discover. What is on the left hand side of the model equation is seldom of primary interest. The pursuit of parameters brought about a puzzling duality in the customary modeling process. While parameters are traditionally estimated by one kind consideration (say, by minimizing the sum of squares or by maximizing likelihood) the fit of the model to data is judged in the light of some other considerations (say, r^2, χ^2, Akaike's Information Criterion, etc.). If the aim is to have a good fit why not estimate parameters by maximizing the chosen goodness-of-fit measure?

In contrast to the tradition in which the parameters are the focus of inquiry, here the aim is to get good estimates of $E\{\mu\}$ and $\sigma\{\mu\}$ for a multitude of populations. Thus, it is not the parameters of the model equation that are of interest, it is the quality of the estimate on the left hand side of the model equation that matters. This change in focus from the parameters to the $E\{\mu\}$ has consequences for modeling. Because it is the residuals that tell how well the $E\{\mu\}$'s are estimated, whatever reduces the residuals and makes them unbiased improves the estimates of the $E\{\mu\}$'s. The implications of this shift in focus for modeling seem to be, as yet, largely unexplored.

[2] An accessible review of the strength and weaknesses is in Schermelleh-Engel et al. (2003). For the history of least squares see Stigler (1986, Chap. 1); for that of the likelihood idea see Edwards (1974).

[3] The weight should be the reciprocal value of the variance of the observed value (see e.g., Aitken 1935; Cameron and Trivedi 1998, p. 28). In Chap. 6 the reciprocal of the fitted value served as weight. This might be reasonable if one assumed that the fitted value is an estimate of the μ of the unit and the observed value for that unit came from a Poisson distribution with that μ as mean. But the fitted value is an estimate of $E\{\mu\}$, not of μ. To do justice to this circumstance one might have to further assume that the μ comes from a Gamma distribution and therefore observed value come from a Negative Binomial distribution. But, at this stage of modeling, the variance of the assumed Gamma distribution is not available. Besides, too many assumptions are already piling up. One might as well make the assumptions explicit in a likelihood equation and estimate the parameters by maximizing likelihood.

will be taken to explain it in some detail.[4] The concepts of "likelihood" will be introduced in Sect. 8.2. A few commonly used "likelihood functions" will be the subject of Sect. 8.3.

The dominance of likelihood maximization as the rationale in parameter estimation stems from the historical interest of scientists and statisticians in the value of parameters. However, when the SPF is to provide information needed in practical road safety management, the focus shifts from the parameters on the right hand side of the model equation to the estimates of $E\{\mu\}$ and $\sigma\{\mu\}$ on its left hand side. This shift in focus affects a change in the objective of optimization. Now that the aim is to get good estimates of $E\{\mu\}$ and $\sigma\{\mu\}$ alternative objective function may be appropriate; these will be examined in Sect. 8.4.

8.2 Likelihood

In a dictionary "likelihood" is a synonym for "probability"; not so in statistics. To explain the difference suppose that in 4 consecutive years 1, 7, 4, and 0 accidents occurred on a road segment with unchanging traits. Assuming that accident counts are Poisson distributed[5] and that the number of accidents occurring in 1 year has no influence on the number occurring in another year, the probability of 1, 7, 4, and 0 accidents to occur is $P(K=1\&7\&4\&0|\mu)=P(K=1|\mu)\times P(K=7|\mu)\times P(K=4|\mu)\times P(K=0|\mu)=\mu^{12}\ e^{-4\mu}/(1!7!4!0!)$. This expression is a function of μ and when viewed as such is called the "likelihood function." The function is shown in Fig. 8.1.

In what follows $\mathcal{L}(.)$ will be used to denote a likelihood function. The dot in the parenthesis is a placeholder for parameters. Thus, for example, $\mathcal{L}(\mu)$ is the likelihood function of μ.

At first glance the likelihood function looks like a probability density function (pdf) but the two must not be confused. In a pdf the ordinate measures the probability of values on the abscissa to occur. In Fig. 8.1 the ordinate is the probability of 1, 7, 4, and 0 accidents to occur if the corresponding parameter value on the abscissa was

[4] What detail is provided should be dictated by what the reader may not already know. This, however, is a difficult judgment to make. Some readers may be bored if the text deals with what to them is well known; others may be frustrated by the unfamiliar and the unexplained. In the "Preface" I set myself the goal "to promote the understanding that is at the core of good modeling". With this in mind I will try to explain the concepts sufficiently and simply. ("What can be said at all can be said clearly; and whereof one cannot speak thereof one must be silent." – Ludwig Wittgenstein). I will tell what assumptions are being made and what their consequences are. For this, reliance on probability and the use of mathematics is unavoidable. The details and derivations have been sequestered into the appendix. Readers who find the going tough may want to engage in judicious skipping.

[5] Let "K" stand for "count of accidents" and "k" for some specific value thereof. By the Poisson distribution the probability to observe k accidents when the mean is μ is $P(K=k)|\mu) \equiv P(k)=\frac{\mu^k e^{-\mu}}{k!}$. This probability law arises when events (here accidents) are rare and their occurrences are independent. More about this is in Appendix A.

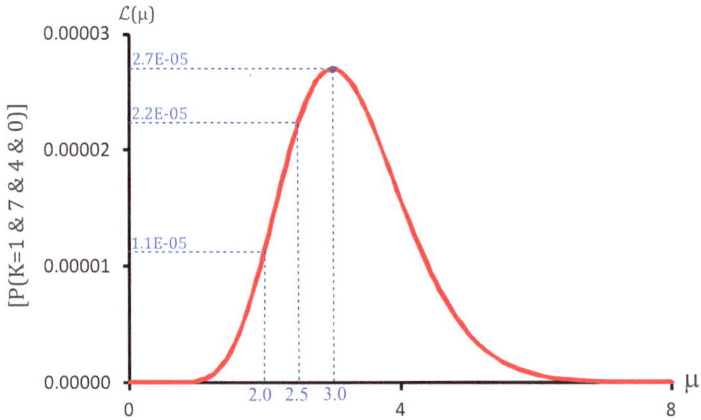

Fig. 8.1 Likelihood function

true. Thus, for example, the 1.1E – 05 does not say how probable $\mu = 2$ is, it says how probable it would be to observe the counts 1, 7, 4, and 0 if μ was 2.

While in view of the 1, 7, 4, and 0 accident counts the $\mu = 2$ is less "likely" than, say, the $\mu = 2.5$, this does not mean that $\mu = 2$ is less probable. For all we know in some population of road segments there may be more segments with $\mu \cong 2.0$ than with $\mu \cong 2.5$ and if so, $\mu \cong 2$ would be more probable.

The peak of the likelihood function is that value of μ which is best supported by the observed data. The $\mu = 3.0$ in Fig. 8.1 is the "maximum likelihood (ML) estimate" of μ.[6] In current SPF development practice the most common method of estimating parameters is that of finding values that correspond to the peak of the likelihood function.

To compute the likelihoods in Fig. 8.1 an assumption had to be made about the probability distribution of accident counts; the assumption was that they are Poisson distributed. More generally, whenever parameters are to be estimated by ML one has to be able to write the likelihood function expression. To do so, one has to be explicit about what is assumed to be the probability distribution of the observed values in which the sought parameters play a role. This is the distinctive feature of maximum likelihood approach to parameter estimation.[7]

---

[6] The maximum likelihood estimate is not the most probable value of the parameter; it is that value of the parameter which makes the observed values most probable. It is in this sense that the parameter value of 3 is best supported by the data. To go to the next step and say which parameter value is most probable one would have to have some prior information about the probability of parameters and a foot in the Bayesian camp of statisticians.

[7] For ordinary least squares no assumption needs to be made about the distribution of observed values. However, one still has to be able to support the assumption that all variances are equal. The ML and the least squares approaches to parameter estimation are equivalent when the model equation is linear and the observed variables are normally distributed.

Whenever a likelihood function is used one has to be explicit about the assumptions made and then ask whether these correspond to what is known.

Two simple examples will illustrate the main features of ML estimation on a spreadsheet. The examples will focus on estimating the parameters of the Poisson and of the Negative Binomial (NB) distribution, both featuring later in SPF development.

8.2.1 The Parameter Behind Poisson Accident Counts

In this example the task is to use a spreadsheet for finding that μ at which the likelihood in Fig. 8.1 is largest and to determine the accuracy of this ML estimate. The four accident counts are in column B of Fig. 8.2 and the corresponding Poisson probabilities[8] are in column C. These depend on the parameter μ in C2. The product of the four probabilities, the likelihood, is in C8. This is the "Target" cell for the Solver to maximize "by changing" the initial guess in C2. Solver finds the ML estimate of μ to be 3.00.[9]

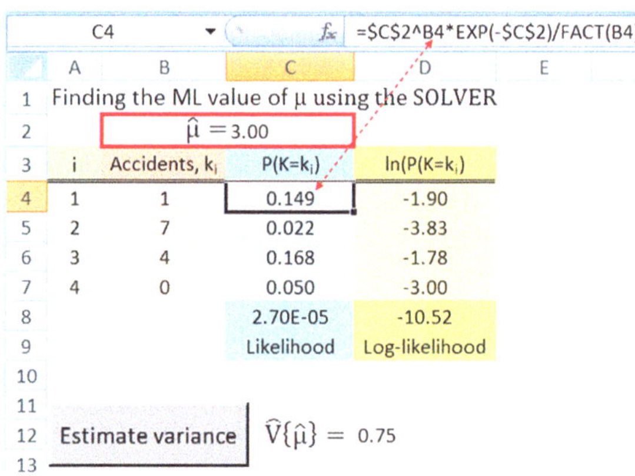

C4 =$C$2^B4*EXP(-$C$2)/FACT(B4)

	A	B	C	D	E
1	Finding the ML value of μ using the SOLVER				
2			$\hat{\mu}$ = 3.00		
3	i	Accidents, k_i	P(K=k_i)	ln(P(K=k_i))	
4	1	1	0.149	-1.90	
5	2	7	0.022	-3.83	
6	3	4	0.168	-1.78	
7	4	0	0.050	-3.00	
8			2.70E-05	-10.52	
9			Likelihood	Log-likelihood	
10					
11					
12	Estimate variance		$\hat{V}\{\hat{\mu}\}$ = 0.75		
13					

Fig. 8.2 Finding the ML estimate μ for Poisson data

[8] The Excel function POISSON(*x*, mean, cumulative) could be used. To get the value in cell C4 one would enter "=POISSON(B4,$C$2,False)."

[9] To download this spreadsheet go to http://extras.springer.com/ and enter the ISBN of this book. Look in the "Spreadsheets" folder for "Chap. 8. Poisson Likelihood. xls or xlsx."

Note that the sample mean (1 + 7 + 4 + 0)/4 is also 3.00. That the sample mean and the ML estimate of μ are the same is not a happenstance.[10] The sample mean is that statistic which makes the expected squared distance to the true mean smallest. It turned out here that the same statistic also maximizes likelihood. This can be generalized further saying that when the observations come from an exponential family and mild conditions are satisfied, least squares and maximum likelihood estimates are identical (Charnes et al. 1976).

Column D in Fig. 8.2 contains the natural logarithm of the values in column C. It was added for two reasons.[11] First, because the likelihood (in C8) is the product of several numbers and the danger is that were more accidents counts used this product might become too small to be represented in a computer.[12] Logarithmization converts multiplication into addition and thereby surmounts this computational obstacle while preserving the location of the maximum.[13]

The second reason for adding column D is that the log-likelihood is needed to compute the statistical accuracy of ML parameter estimates.[14] The numerical

---

[10] When accident counts are Poisson distributed the likelihood function is $\mathcal{L}(\mu) = \mu^{\sum_1^n k_i} e^{-n\mu} / \Pi_1^n k_i$ where $k_1, k_2, \ldots, k_n$ are the accident counts. The derivative is $d\mathcal{L}(\mu)/d\mu = \mathcal{L}(\mu)(\sum_1^n k_i/\mu - n)$. At the peak the derivative is 0. This occurs when $\sum_1^n k_i/\mu - n = 0$. It follows that the ML estimate of μ is the same sample mean and using the Solver, in this case, was unnecessary.

[11] There is a third reason for using the log of the likelihood function and it has to do with determining whether the addition of a parameter to the model equation is justified. This issue will be discussed in Chap. 10.

[12] The smallest allowed positive number is 2.2251E − 308. This limit would be exceeded if 300 or so accident counts were used. In the Colorado data, e.g. there are 5,323 accident counts.

[13] The logarithmic transformation is "monotonic," i.e., order preserving. That is, if the likelihood when $\mu = 2.0$ is smaller than when $\mu = 2.5$ (as in Fig. 8.1) then the log of the likelihood when $\mu = 2.0$ is smaller than the log-likelihood at $\mu = 2.5$. This is why $\mathcal{L}(\mu)$ and $\log(\mathcal{L}(\mu))$ have a peak over the same μ. Along the same lines, instead of using the Poisson probability $\frac{\mu^k e^{-\mu}}{k!}$ in column C one could omit the $k!$ and use the "abridged" likelihood $\mu^k e^{-\mu}$ or the "abridged" log-likelihood $k\ln(\mu) - \mu$ in column D. This too is a monotonic transformation and avoids problem of Excel being unable to return factorials larger than 170.

[14] When μ is estimated maximizing likelihood then $V\{\hat{\mu}\}$ is a function of the second order partial derivatives of the log-likelihood. For a single parameter the variance of the ML estimate is the reciprocal of the negative expected value Fisher Information (Geyer, 2003). The Fisher Information is the second partial derivative of the log-likelihood. For the Poisson likelihood $\mathcal{L}(\mu) = \mu^{\sum_1^n k_i} e^{-n\mu} / \Pi_1^n k_i$ the first partial derivative is $\frac{\partial \ln \mathcal{L}(\mu)}{\partial \mu} = \frac{\sum_1^n k_i}{\mu} - n$ and the second partial derivative is $\frac{\partial^2 \ln \mathcal{L}(\mu)}{\partial \mu^2} = -\frac{\sum_1^n k_i}{\mu^2}$. The Fisher Information is $-E\left\{\frac{\sum_1^n k_i}{\mu^2}\right\}$, its observed value is $\frac{\sum_1^n k_i}{\left(\sum_1^n k_i/n\right)^2} = \frac{n^2}{\sum_1^n k_i} = \frac{n}{\text{Sample Mean}}$ and therefore the variance of $\hat{\mu}$ is estimated as Sample Mean/n. This, again, is not a surprise. Because the least squares and the maximum likelihood estimates of μ are identical, their variances are also the same; here the sample mean is 3, $n = 4$ and $V\{\hat{\mu}\} = \frac{3}{4} = 0.75$. The procedure requires the determination of the second order partial derivative of the log-likelihood. Approximating this derivative by finite differences the corresponding numerical expression is $\frac{\partial^2 \ln \mathcal{L}(\mu)}{\partial \mu^2} \cong \frac{\ln \mathcal{L}(\hat{\mu} + \Delta) - 2\ln \mathcal{L}(\hat{\mu}) + \ln \mathcal{L}(\hat{\mu} - \Delta)}{\Delta^2}$ where Δ is, say, $\hat{\mu}/100$.

procedure for computing the estimate of $V\{\hat{\mu}\}$ has been coded into the command button "Estimate variance" in Fig. 8.2 and the result is in cell D12.

Several features of this simple example are of general significance and will be used whenever parameters are estimated by maximizing likelihood. First, instead of using the product of probabilities the sum of their natural logarithms will be maximized. Second, that the order preserving abridged log-likelihood function will be used.[15] Third, that the statistical accuracy of parameter estimates will be determined either by a numerical procedure or by simulation.

8.2.2 The Parameters Behind the NB Distribution

Data about the number of Connecticut drivers with "k" accidents in 1931–1936 were introduced earlier in Table 1.3 and are repeated here in columns A and B of Fig. 8.3. Thus, for example, 4,503 drivers out of a total of 29,531 had 1 accident during the 1931–1936 period. Using these, the parameters "a" and "b" of the Gamma and the NB distribution were estimated earlier by the "Method of Moments."[16] Now, in Fig. 8.3, the same parameters are estimated by maximizing likelihood.

When the number of accidents for a driver is Poisson distributed with a mean of μ and this μ comes from a population in which the μ's are Gamma distributed with $E\{\mu\} = b/a$ and $V\{\mu\} = b/a^2$ then, as shown in Appendix D, the probability of a driver of this population to have "k" accidents is given by the NB distribution:

$$P(k) = \frac{\Gamma(k+b)}{\Gamma(b)k!}\frac{a^b}{(a+1)^{k+b}} \tag{8.1}$$

The natural logarithm of $P(k)$ is in cells C5:C12.[17] The probability of $n(k)$ drivers to each record "k" accidents is $P(k)^{n(k)}$ and the natural logarithm thereof is in cells D5:D12.

The log-likelihood for all drivers is in D13; this is the "Target" cell for Solver. The "By Changing" cells are D2:E2. The Solver returns the ML estimates of "a" and "b" in Fig. 8.3. These differ somewhat from those obtained by the method of moments.[18] The fitted values in column E are the number of drivers expected to have k accidents if the ML estimates of "a" and "b" were the parameters. The correspondence is extraordinary.

[15] While $\mathcal{L}$ was used to denote the likelihood function, $\mathcal{L}^*$ will denote the abridged likelihood function.

[16] See (1.4). The Method of Moments introduced by Karl Pearson in 1984 is one of two general approaches to parameter estimation. The other approach is that of maximizing likelihood.

[17] The formula in C5 is "=$E$2*LN($D$2) – GAMMALN($E$2) – LN(FACT(A5)) + GAMMALN(A5 + $E$2) – (A5 + $E$2)*LN($D$2 + 1)".

[18] There the estimates were 3.647 and 0.875.

Fig. 8.3 Finding ML estimates of a and b for Negative Binomial data.[19]

C5 =$E$2*LN($D$2)-GAMMALN($E$2)-LN(FACT

	A	B	C	D	E
1	**Connectic**	3.674623		**a**	**b**
2		0.881234		**3.749**	**0.890**
3					
4	**k**	**n(k)**	**ln[P(k)]**	**n(k)ln[P(k)]**	**Fitted**
5	0	23881	-0.210	-5026.4	23926
6	1	4503	-1.885	-8487.1	4485
7	2	936	-3.499	-3275.2	893
8	3	160	-5.094	-815.1	181
9	4	33	-6.680	-220.4	37
10	5	14	-8.260	-115.6	8
11	6	3	-9.837	-29.5	2
12	7	1	-11.410	-11.4	0
13	**Sums**	**29531**		**-17980.8**	
14				**Log-likelihood**	
15					
16		Compute observed Fisher Information Matrix			
17					
18			Observed Fisher Information		
19			a	b	
20		a	391.9	-1659.3	
21		b	-1659.3	7493.6	
22					
23			Variance-Covariance		
24			a	b	
25		a	0.041	0.009	
26		b	0.009	0.002	

To estimate the variance of the ML estimates of "a" and "b" one has to compute the inverse of the observed Fisher Information Matrix (Geyer, 2003).[20] The

[19] To download this spreadsheet go to http://extras.springer.com/ and enter the ISBN of this book. Look in the "Spreadsheets" folder for Chap. 8. Negative Binomial Likelihood. xls or xlsx.

[20] The Fisher Information Matrix is $\begin{bmatrix} -E\left\{\dfrac{\partial^2 \ln\mathcal{L}(a,b)}{\partial a\,\partial a}\right\} & -E\left\{\dfrac{\partial^2 \ln\mathcal{L}(a,b)}{\partial a\,\partial b}\right\} \\ -E\left\{\dfrac{\partial^2 \ln\mathcal{L}(a,b)}{\partial a\,\partial b}\right\} & -E\left\{\dfrac{\partial^2 \ln\mathcal{L}(a,b)}{\partial b\,\partial b}\right\} \end{bmatrix}$. For the observed matrix the expectation operator (E) is omitted and the second order partial derivatives evaluated at the ML estimates of a and b. The finite difference expression for the derivatives is

$$\frac{\partial^2 \ln\mathcal{L}(a,b)}{\partial a\,\partial b} \cong \frac{\ln\mathcal{L}(a+\Delta a, b+\Delta b) - \ln\mathcal{L}(a+\Delta a, b-\Delta b) - \ln\mathcal{L}(a-\Delta a, b+\Delta b) + \ln\mathcal{L}(a-\Delta a, b-\Delta b)}{4\Delta a\Delta b}.$$

The nature of this unintuitive procedure can be perhaps explained with reference to Fig. 6.6. Minimization and maximization are nearly identical twins. Thus, e.g., one would find the same extremum if instead of minimizing wSSD one would maximize its negative value. This leads to the following reasoning: Suppose that a parameter is estimated by maximizing likelihood or, equivalently, by minimizing negative likelihood. The first derivative of the negative likelihood at the minimum is of course 0. It is the second (partial) derivative there that measures the "sharpness of definition" for the minimum. The larger the second derivative, the faster the curve in Fig. 6.6 turns

numerical procedure has been coded into the command button "Compute observed Information Matrix" in Fig. 8.3 and the results are in cell C20:D21. The inverse is the variance-covariance matrix of the parameter estimates in range C25:D26.[21]

Having introduced the notion of likelihood function and the estimation of parameters by its maximization, the next task is to prepare a few likelihood functions to be used for SPF curve-fitting.

8.3 A Few Likelihood Functions

The Poisson and the NB likelihood functions will be introduced first. Both are commonly used in SPF development. These will be followed by a third likelihood function, the Negative Multinomial (NM) which makes better use of available data. Details about the assumptions behind these likelihood functions and their derivation are in the Appendix.

8.3.1 The Poisson Likelihood Function

The problem with minimizing the sum of squared differences in Chap. 6 was that while the variance of the accident counts should have been constant, it clearly differed from segment to segment. An observed value with a small variance should carry more weight than one with a large variance but the question of what weights should be used was left without a satisfactory answer.

To alleviate the unequal variances problem, in this section accident counts are assumed to come from a Poisson distribution. For the Poisson distribution mean = variance, and since the mean ($E\{\mu\}$) increases with Segment Length, so does the variance of the accident counts. In this manner the question of what weight to use is automatically dealt with.

To state the Poisson likelihood function one assumes (i) that the accident counts on each unit are Poisson distributed, (ii) that units with the same traits in the model equation have the same μ, and (iii) that the accident count on one unit is statistically independent of the accident count on another unit. As shown in Appendix B, with data about n units the (abridged) log-likelihood function is:

$$\ln[\mathcal{L}^*(\mu_1, \ldots, \mu_n)] = \sum_{i=1}^{n} (-\mu_i) + k_i \ln(\mu_i) \tag{8.2}$$

upward, the narrower is the flat plateau, the sharper is definition of the minimum, and the lesser is the variance of the parameter estimate. From here it is just a short step to the "Fisher Information" which is made of the negative second order partial derivatives of the negative log-likelihood.

[21] The Excel function MINVERSE for matrix inversion was used.

Replacing the μ's by the model equation $\hat{E}\{\mu_i\} = \beta_0(\text{Segment Length}_i)^{\beta_1}$ turns the log-likelihood into a function of β_0 and β_1, i.e., into $\ln[\mathcal{L}^*(\beta_0, \beta_1)]$. The fitting of this model equation to data amounts to finding those values of β_0 and β_1 that maximize $\ln[\mathcal{L}^*(\beta_0, \beta_1)]$.

Implementation on the C-F spreadsheet is simple.[22] The spreadsheet in Fig. 8.4 is the same as that in Fig. 6.3 except that instead of minimizing the sum of squared differences here the abridged log-likelihood is maximized. Accordingly, the SD in column E of Fig. 6.3 is replaced by the abridged log-likelihood from (8.2).[23]

The "Target" cell for the Solver, the "by Changing" cells, and the constraint all remain the same. Only the radio button in the Solver window has to be changed from "Min" (which is appropriate for minimizing sum of squared differences) to "Max" (for maximizing log-likelihood.) The ML estimate of β_1 is now 0.860, nearly the same as in Figs. 6.3 and 6.4 where squared differences were minimized. Because the parameters are very similar, their CURE plots are bound to be similar.

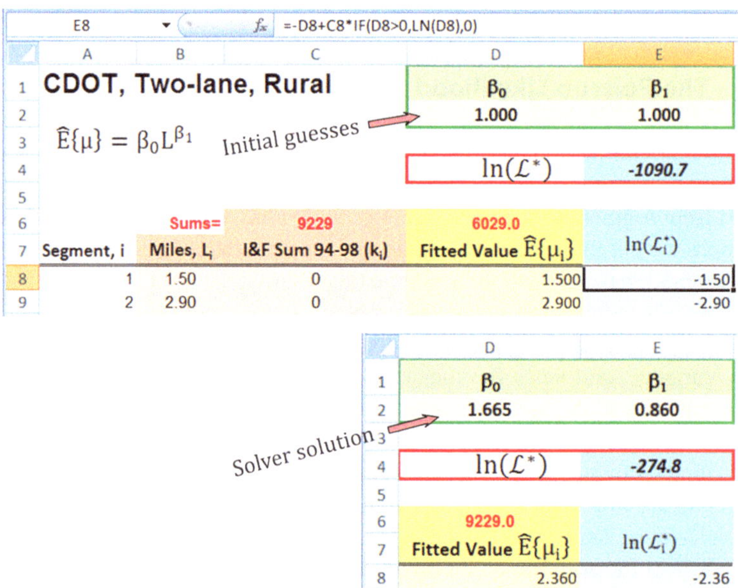

Fig. 8.4 Poisson ML fit to the condensed Colorado data

[22] To download this spreadsheet go to http://extras.springer.com/ and enter the ISBN of this book. Look in the "Spreadsheets" folder for "Chap. 8. Poisson Likelihood fit. xls or xlsx."

[23] The formula in E8 is D8 + C8*IF(D8 > 0,LN(D8),0). The condition is there because the natural logarithm is defined only for positive numbers and the Solver might choose a step resulting in a negative value for β_0. Whenever the LN function is invoked it is prudent to use the appropriate "IF" condition. An alternative is to add to the Solver the constraint $\beta_0 > 0$.

In most cases the variance = mean property of the Poisson distribution is also not supported by data. To illustrate, amongst the 5,323 Colorado road segments, there are 91 segments that are all 0.01 miles long. If they all had the same μ then the variance of the accident counts on these segments should be also μ. But this is not what the data say. The sample mean of accident counts for these segments is 0.099 while the sample variance is 0.112. The results of similar computations for 50 Segment Length bins are in Fig. 8.5. Almost all variance-to-mean ratios estimates exceed 1. This is an indication that in this data set accident counts are "overdispersed." Overdispersion will occur if on each unit accidents are Poisson distributed but segments that are identical in length do not have identical μ's.

That segments identical in length differ in their μ's is obvious.[24] Thus, for example, amongst those 91 segments which are all 0.01 miles long the AADTs range from 300 to 14,000 and one must expect that the μ depends on AADT. Generalizing, one usually finds that in populations of units defined by a limited number of traits, the sample variance of accident counts is larger than their sample mean. This is a sign that the $\sigma\{\mu\} > 0$. Because in the Poisson likelihood function the variability amongst μ's can neither be accommodated nor estimated, a different probability distribution must be sought.

As noted earlier, all likelihood functions are based on assumptions. For ML estimation to be convincing, the assumptions made should be in line with logical considerations and empirical facts. The Poisson likelihood function does not meet this requirement. Neither by logic nor by data can one justify the assumption that units identical in the few traits that are in the model equation all have the same μ. The NB likelihood function allows the μ's to differ and thereby meets this requirement, at least in part.

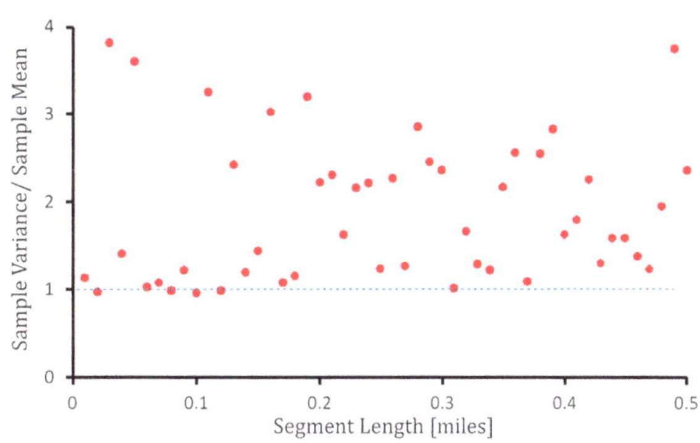

Fig. 8.5 Estimates of variance/mean for 50 Segment Length bins

[24] The logical objection to assuming that for a population of units with the same few traits $\sigma\{\mu\} = 0$ was stated in Sect. 1.2 as follows: "...for any population of units, no matter how many traits are used to define it, the only realistic point of departure is to assume that the μ's of its units are all different."

8.3.2 The Negative Binomial Likelihood Function

The NB distribution was already mentioned and used with the Connecticut drivers and the Colorado road segments data. It is based on four assumptions: (i) that, as before, the accident counts on every unit are Poisson distributed, (ii) that the μ's of the units in each population (real or imagined) are different, i.e., that $V\{\mu\} > 0$, (iii) that the diversity of the μ's in each population can be well approximated by a Gamma distribution and (iv) that the "shape" parameter of the gamma distributions for all populations behaves in a specific manner. More about this is in Appendices D and E.

For assumption (i) one can muster some logical support.[25] Assumption (ii) follows from the foundational axiom in Sect. 1.2. However, there is no logical support for assumptions (iii) and (iv). Some data support assumption (iii), see e.g., Table 1.6, while other data negate it, see, e.g., Table 1.7. The common use of the Gamma assumption is mostly a matter of convenience: if (i) and (iii) hold then the probability to observe k accidents is Negative Binomial and the expression for the likelihood function can be written. It follows that before the NB likelihood function is used for parameter estimation one may want to check whether the data do not negate the Gamma assumption.

As shown in Appendix E the abridged NB log-likelihood is:

$$\ln[\mathcal{L}^*(\beta_0, \beta_1, \ldots, \ell)] = \sum_{i=1}^{n} \begin{bmatrix} \ln\Gamma(k_i + b_i) - \ln\Gamma(b_i) + b_i\ln(b_i) + k_i\ln\left(\widehat{E}\{\mu_i\}\right) \\ -(b_i + k_i)\ln\left(b_i + \widehat{E}\{\mu_i\}\right) \end{bmatrix} \tag{8.3}$$

As in Sect. 8.1 earlier, $\hat{E}\{\mu_i\}$ is replaced by the model equation which is a function of the parameters $\beta_0, \beta_1, \ldots$. In addition, assumption (iv) takes the specific form[26] of $b_i = \ell L_i$. The $\ln\Gamma(.)$ in (8.3) is the natural logarithm of the Gamma function.[27] Implementation in the C-F spreadsheet of Fig. 8.6 amounts to coding (8.3) into E8 and the cells below.[28]

The ML estimate of β_1 is 0.871, close to the 0.860 obtained earlier by the Poisson fit. The new per-unit-length overdispersion parameter (ℓ) is in F2. To illustrate its use consider a 0.7 miles long road segment. The estimate of the $E\{\mu_i\}$ for a population of road segments of this length is $1.656 \times 0.7^{0.871} = 1.2$ I&F accidents

---

[25] See Appendix A.

[26] The reciprocal value of ℓL_i is the "overdispersion parameter." Once $E\{\mu\}$ is estimated by the model equation the variance of μ follows from $V\{\mu\} = (E\{\mu\})^2/(\ell L_i)$. The larger is "$\ell$" the smaller is $V\{\mu\}$ and the lesser is the "overdispersion" in comparison to the Poisson.

[27] The Gamma function is an extension of the factorial from integers to real numbers When n is an integer. $\Gamma(n) = (n-1)!$. In Excel one can use the GAMMALN(.) function.

[28] To download this spreadsheet go to http://extras.springer.com/ and enter the ISBN of this book. Look in the "Spreadsheets" folder for "Chap. 8. Negative Binomial Likelihood Fit .xls or xlsx."

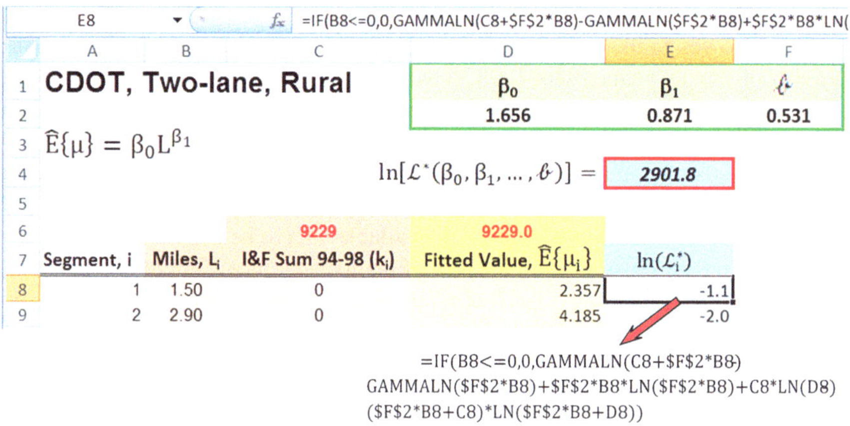

Fig. 8.6 Negative Binomial ML fit to the condensed Colorado data

in 5 years. Because $V\{\mu_i\} = (E\{\mu_i\})^2/b_i$ and $b_i = b \times L_i$, the estimate of the variance of μ's amongst 0.7 miles long road segments is $1.2^2/(0.531 \times 0.7) = 3.87$ (I&F accidents in 5 years)^2. This information is needed for empirical Bayes safety estimation.

The NB likelihood function has two main figures of merit. First, it is consistent with the mental construct of populations of units defined by the traits in the model equation. Because of this consistency, the interpretation of the results of curve-fitting is unambiguous; what we get are estimates of the mean and standard deviation of μ's for each population. Second, because at the core of the NB distribution is the assumption that the μ's a Gamma distributed, the ML estimate of the overdispersion parameter provides a direct way to estimate the $V\{\mu\}$ for all combination of traits and levels. The weakness of this likelihood function is in assumptions (iii) and (iv). There is no assurance that the μ's in any population or subpopulation are approximately Gamma distributed, nor may one assume without support that the overdispersion parameter exhibits the assumed regularity. Convenience in use is a strong temptation but a weak argument.

> The NB distribution is consistent with the image of populations of units with diverse μ's. Each crash count in the data is an observation from a probability distribution, the μ of which is one of the μ's from that population.

8.3.3 The Negative Multinomial Likelihood Function

The Poisson and the NB likelihood function can be used only when the data pertain to a single period of time. This is why, up to this point, the "condensed data" were used and the 5 years 1994–1998 were taken to be a single period. The accident count was the sum of accidents for that 5-year period and AADT was the 5-year average. And yet, in the original data we have accident counts and AADT estimates for each of the 13 years 1986–1998.[29] Not only is the averaging of AADTs over 5 years a stretch, but not using all the available data just because our tools cannot deal with multiple time periods is an unjustified loss of information. This is why when in the next chapter AADT will be added to the variables, the Poisson and NB likelihood function will be replaced by the NM (Negative Multinomial) likelihood function. Not only will it allow all of the available data to be used, it will also bring out and make explicit an essential aspect of reality: that the safety of units changes with time. For completeness, the NM likelihood function is best introduced here.

Consider a sample of n units such that information about unit i is available for time periods $1, 2, \ldots, j, \ldots, m_i$. Data of this kind are called "panel" or "longitudinal." One data element for unit i are the accident counts $k_{i,1}, k_{i,2}, \ldots, k_{i,j}, \ldots, k_{i,m_i}$. As before, accident counts are assumed to be Poisson distributed only now the means are $\mu_{i,1}, \mu_{i,2}, \ldots, \mu_{i,j}, \ldots, \mu_{i,m_i}$, etc. The traits of unit "i" in period "j" define a population of units. The μ's of this population are again assumed to be Gamma distributed. These assumptions lead to an extension of the Negative Binomial to panel data: the Negative Multinomial distribution. As is shown in Appendix G the contribution of unit i to the abridged log-likelihood is:

$$\ln\left[\mathcal{L}_{\mathrm{i}}^{*}(\beta_0, \beta_1, \ldots, \mathscr{b})\right] = \mathscr{b}L_i \ln(\mathscr{b}L_i) + \left[\sum_{j=1}^{m_i} k_{i,j} \ln\left(\widehat{E}\{\mu_{i,j}\}\right)\right]$$
$$+ \ln\Gamma\left(\sum_{j=1}^{m_i} k_{i,j} + \mathscr{b}L_i\right) - \ln\Gamma(\mathscr{b}L_i) - \left(\sum_{j=1}^{m_i} k_{i,j} + \mathscr{b}L_i\right) \ln\left[\left(\sum_{j=1}^{m_i} \widehat{E}\{\mu_{i,j}\}\right) + \mathscr{b}L_i\right] \tag{8.4}$$

This likelihood function will be used in conjunction with yearly data after AADT will be introduced into the model equation.

8.4 Alternative Objective Functions

Minimizing the sum of weighted squared residuals or maximizing likelihood are the two mainstream approaches to parameter estimation. One reason for their popularity is that researchers are mainly interested in parameters. The parameter, especially when significantly different from 0, is thought to say something about the make-up of the world.[30] With least squares or maximum likelihood approaches to estimation,

[29] See Sect. 3.2.

[30] The research perspective on modeling was discussed in Sect. 1.5. A jaundiced opinion about the research perspective and the prospects of success for the causal interpretation of parameters is in Sect. 6.7.

the variance of parameter estimates can be minimal, statistical hypotheses about the parameters can be tested, and confidence limits stated.

When in SPF modeling the focus is on applications attention shifts from parameters to estimates and predictions of[31] $E\{\mu\}$ and of $V\{\mu\}$. This change of focus opens the door to the use of other objective functions. Now it is not the parameter value and the testability of statistical hypotheses about it that are of interest; now one wants the model to produce good estimates and predictions of $E\{\mu\}$ and $V\{\mu\}$ for use in practice.

If the aim is to get good estimates or predictions for $E\{\mu\}$ and $V\{\mu\}$ why would one want to minimize the sum of squared differences; would it not make more sense to minimize, say, the sum of absolute residuals[32]? Why not minimize the χ^2 or maximize the r^2 of a fit[33] when these are widely used to measure the goodness of a fit? Why not choose for the objective function, one of the many other measures of forecast accuracy that are in use[34]?

A few alternative objective functions were tried and the results are in Fig. 8.7.[35] Their implementation required only small adaptations of the C-F the spreadsheet in Fig. 8.6. Thus, for example, to minimize the sum of χ^2 or of absolute residuals only the formula in column E had to be changed; to maximize r^2 the excel function SQR was used in cell E4.[36]

---

[31] The difference between "estimate" and "prediction" is in their relation to the data set used for SPF development. Thus, e.g., one may estimate the parameters of a model equation using a certain data set, or one may use the model equation to estimate the $E\{\mu\}$ of a certain unit in that data set. In both cases the estimate pertains to that data set. However, when one uses this model equation with its estimated parameters in practice, it is to predict the $E\{\mu\}$ for a unit or for a period that was not in the data set used when the SPF developed.

[32] The choice between maximizing likelihood and minimizing the sum of absolute differences is reminiscent of the more familiar choice between the "sample mean" and the "sample median." Both are used to describe the central tendency in data, both have advantages and drawbacks The sample mean makes the average squared error of predictions smallest; the sample median minimizes the average (absolute) error of predictions. The sample mean is an unbiased and minimum-variance estimate of the population mean but is sensitive to outliers and to randomness in skewed distributions. The sample median is a more robust estimator, less influenced by errors and deviant observation.

[33] $\chi^2 \equiv$ (Recorded accidents − Accidents predicted by the model)^2/(Accidents predicted by the model) summed over all data points.

[34] Common measures of forecast accuracy are the MAPE (Mean Absolute Percentage Error), MdAPE (Median Absolute Percentage Error), sMAPE (Symmetric Mean Absolute Percentage Error), sMdAPE (Symmetric Median Absolute Percentage Error), MdRAE (Median Relative Absolute Error), GMRAE (Geometric Mean Relative Absolute Error), and the MASE (Mean Absolute Scaled Error). From Hyndman and Koehler (2006).

[35] To download this spreadsheet go to http://extras.springer.com/ and enter the ISBN of this book. Look in the "Spreadsheets" folder for Chap. 8. Unconventional objective functions. xls or xlsx.

[36] The Pearson product moment coefficient of correlation, r, is $\frac{\Sigma_1^n (X_i-\bar{X})(Y_i-\bar{Y})}{\sqrt{\Sigma_1^n (X_i-\bar{X})^2}\sqrt{\Sigma_1^n (Y_i-\bar{Y})^2}}$; its square is computed by the Excel function SQR(...). Here the dots stand for C8:C5330 and D8:D5330.

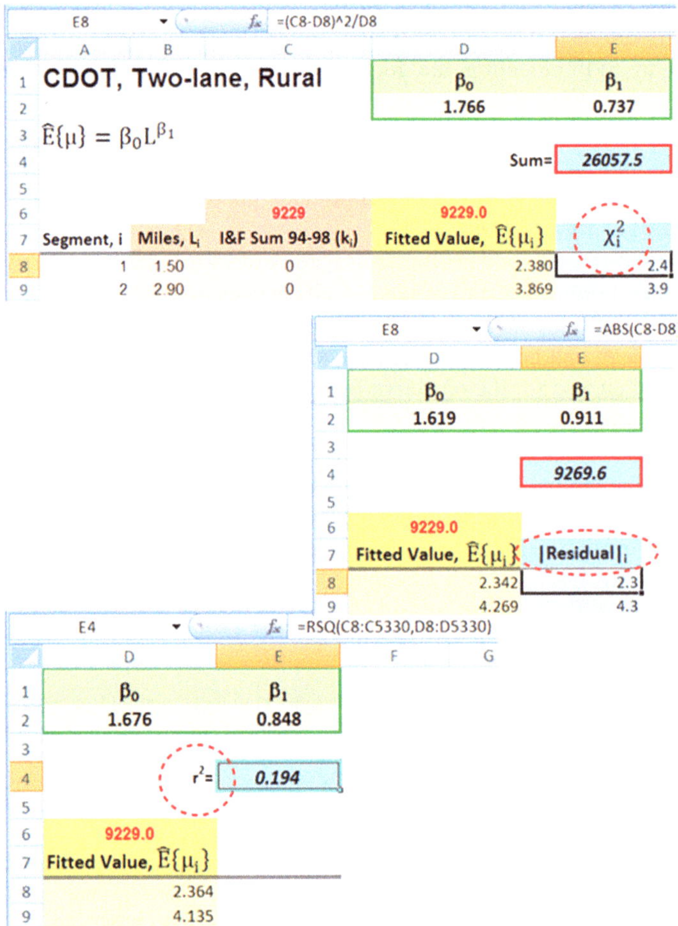

Fig. 8.7 Parameter estimates when sum of χ^2 and sum of |residuals| are minimized, and when r^2 is maximized

The CURE plots corresponding to these objective functions are compared to that of minimizing the weighted sum of squared differences in Fig. 8.8.[37] While the minimization of χ^2 is seen to be inferior to the wSSD, maximizing r^2 yields essentially the same fit quality, and the minimization of absolute residuals is better.

However, even this best fit is not satisfactory. All four CURE plots show regions of large bias-in-fit. Choosing one objective function or another just moves the CURE plots up or down between the fixed endpoints leaving their in-between shape largely unaffected. To affect this shape and to improve the fit the model

[37] The CURE plot for wSSD represents also the Poisson and NB likelihood CURE plot which had very similar parameter estimates.

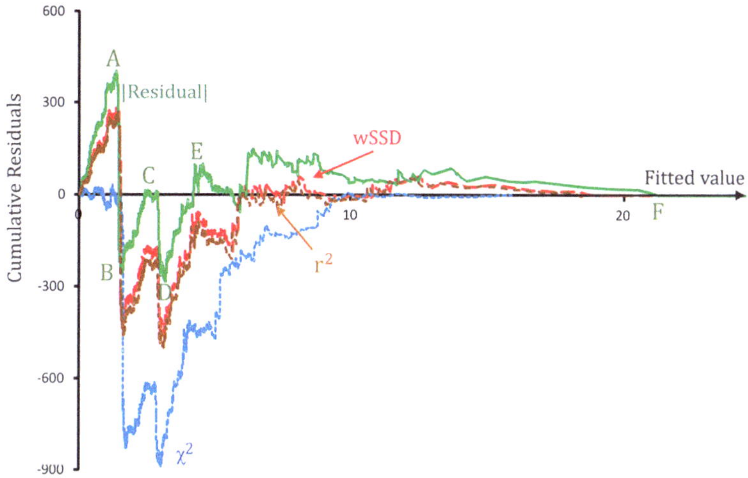

Fig. 8.8 Comparing CURE plots

equation has to be changed. This can be done by choosing a more suitable function and by adding variables.

Yet another alternative objective function merits attention. How to compute the bias-in-fit was shown in Sect. 7.3. Inasmuch as a good SPF model is one that gives unbiased estimates of $E\{\mu\}$ at all variable values of practical interest the minimization of the Total Absolute Bias is a potentially interesting objective function. The corresponding C-F spreadsheet is in Fig. 8.9.

	D	E	F	G
1	β_0	β_1		
2	1.637	0.892		
3				
4	TAB=	1752.5		
5			Cumulated	
6	9229.0		Accidents	Fitted
7	Fitted Value	Polygon		
8	0.027		0.0	0.0
9	0.027		0.0	0.1

K4 f_x =F2546

	I	J	K	L	M	N
1	Polygon		Cumulated			
2	Points	Row	Accidents	Fitted values	Residual	|Bias|
3	O		0	0		
4	A	2546	1751	1410	341	341
5	B	3571	2649	2930	-281	622
6	C	4346	4446	4515	-69	213
7	D	4658	5114	5437	-323	254
8	E	4946	6532	6532	0	323
9	F	5330	9229	9229	0	0
10					Sum=	1753

Fig. 8.9 Minimizing the total accumulated bias

The letters A to F correspond to the high and low points of the CURE plot in Fig. 8.8. Thus, for example, point A is in row 2,546 where, in column F the accumulated accident count is 1,751 and, in cell G2546 the number of accumulated fitted accidents is 1,410. Thus, between the origin O and point A the excess of residuals is 341. Similarly at point B the excess of residuals is −281 and the bias between A and B is 341 − (−281) = 622 accidents. Proceeding similarly, the total accumulated bias is in N10. Because the numbers in columns K and L depend on the parameters in D2:E2, so does the total accumulated bias. The parameter estimates in Fig. 8.9 are those that minimize the TAB; they are similar to those that minimize the sum of absolute residuals.

A few observations follow. First, alternative objective functions deserve attention. They are simple to implement and can produce fits that are as just as good or better than those obtained by least squares or by maximum likelihood.

Second, when the purpose of SPF modeling is to produce good estimates of $E\{\mu\}$ and $\sigma\{\mu\}$ the use of an alternative objective function derives directly from the purpose of modeling. Eliminated is the duality and tension characterizing least squares and maximum likelihood approaches to estimation where parameters are estimated with one aim in mind and goodness-of-fit measured with a different one.

Third, the use of a spreadsheet has a liberating influence on modeling. No statistical software now in existence can estimate parameters that minimize, say, Total Absolute Bias, the area under the CURE plot or similar alternative objective functions. But a creative modeler can easily produce a corresponding spreadsheet.

Fourth, modelers bound to the conventions of least squares or maximum likelihood parameter estimation will worry about conditions that may not hold and assumptions made that may or may not be approximately true. Are the variances of the observed values equal? If not, what weights should be used? Do the μ's well approximate the Gamma distribution in all populations? Is the overdispersion parameter a constant? What deviation from the necessary assumptions compromises modeling results? Questions of this kind simply do not arise with the alternative objective functions. If χ^2, r^2, or |residuals| are thought to measure goodness-of-fit then these are the values to minimize or maximize. If the aim is to bring the curve or surface as close to the data as possible then minimizing the sum of absolute residuals makes sense; there are no unmet conditions nor assumptions that may not be justified.

Fifth, as shown in Table 8.1 parameter estimates depend to some extent on the objective function chosen. This variability has nothing to do with the "statistical inaccuracy."[38] Rather it illustrates its incompleteness due to neglect of other sources of uncertainty.[39] Here the uncertainty is about what objective function to use.

---

[38] The statistical inaccuracy was defined and discussed in Sect. 6.6.1.

[39] The other sources of uncertainty were called "modeling inaccuracy" and were discussed in Sect. 6.6.2.

Table 8.1 Objective functions and estimates of β_1

	Objective function	$\hat{\beta}_1$
Conventional	Unweighted least squares	0.866
	Fixed weight least squares	0.859
	Variable weight least squares	0.737
	Poisson likelihood	0.860
	Negative Binomial likelihood	0.871
Alternative	Absolute differences	0.911
	χ^2	0.737
	Total absolute bias	0.892
	r^2	0.848

Which objective function is right, which estimate is more trustworthy? One cannot say. Three of the estimates in Table 8.1 are outside the 95 % confidence limits of $\hat{\beta}_1$ established in Sect. 6.6.1. This underscores that fact that the uncertainty surrounding parameter estimates is not only due to the randomness of the data (which the confidence interval reflects) but, amongst other reasons, due to the uncertainty about what to minimize; a choice that must be made but is not easily defended.

8.5 Summary

When parameters were estimated by least squares the fits were bad. A poor fit can come about for several reasons one of which is the use of an inferior objective function. In this section the focus was on two kinds of objective functions: those based on likelihood and those based on goodness-of-fit.

The concepts of likelihood, the likelihood function, the abridged log-likelihood, and the maximum likelihood estimate were explained and illustrated by numerical examples. They show how the Solver is used to estimate ML parameters and how their statistical accuracy can be determined using Fisher's Information. In preparation for further model development the commonly used Poisson, the NB, and the NM likelihoods were given and their implementation in the C-F spreadsheet was illustrated.

The distinguishing feature of fitting SPFs by maximizing likelihood is that one must make some explicit assumptions. For the Poisson likelihood function one assumes (i) that the accident counts on each unit are Poisson distributed, (ii) that units with the same traits in the model equation have the same μ, and (iii) that the accident count on one unit is statistically independent of the accident count on another unit. With these assumptions the abridged Poisson log-likelihood function could be written. To make it operational, the Poisson mean was replaced by the model equation. Implementation in a C-F spreadsheet was simple.

The assumption that units with the same traits in the model equation all have the same μ does not make good sense and is contrary to empirical evidence. Use of the

NB distribution is a popular remedy. The NB likelihood function retains assumption (i) and (iii) but replaces (ii) by assuming that the μ's of the units of which a population is comprised come from a two-parameter Gamma distribution. With this the abridged log-likelihood function could be written. To make it operational, the mean of the Gamma distribution was replaced by the model equation and an assumption was added about its shape parameter. The implementation on a C-F spreadsheet was again straightforward. The attraction of the NB likelihood function is that it provides estimates of $E\{\mu\}$ and $V\{\mu\}$ at the same time.

The Poisson and the NB likelihood functions cannot be reasonably used when data are about multiple time periods and some traits (e.g., AADT) change from period to period. The Negative Multinomial (NM) is the extension of the NB likelihood function for such circumstances. It shares the assumptions of the NB distribution and assumes, in addition, that the μ's of all units in a population change from year to year just as the mean of these μ's changes.

Much of this section revolved around the concept of likelihood and the discussion of three commonly used likelihood function. The overall topic, however, was the question what to optimize. This question can be answered more generally when the purpose of SPF modeling is not doing research about parameters but meeting the practical needs of road safety management. With this change in focus several alternative objective functions become sensible. The optimization of fit by minimizing χ^2, maximizing r^2, minimizing the sum of |residuals| and of Total Accumulated Bias was examined. The use of alternative objective functions proved both simple and attractive.

What a parameter is estimated to be depends to some extent on the chosen objective function. The choice of the objective function is made by the modeler. It is of course disconcerting that reasonable looking objective functions produce discrepant parameter estimates. One may be guided in the choice of objective function by comparing CURE plots. Of the several objective functions examined here the sum of absolute residuals proved best. But even the best fit was still unsatisfactory. Important variables may be missing from the model equation, the function chosen to serve as the model equation may be unsuitable, and there may be outliers that distort the CURE plot. These issues will be tackled in Chaps. 9 and 10.

References

Aitken AC (1935) On least squares and linear combinations of observations. Proc Roy Soc Edinburgh 55:42–48

Cameron AC, Trivedi PK (1998) Regression analysis of count data. Cambridge University Press, Cambridge

Charnes A, Frome EWL, Yu PL (1976) The equivalence of generalized least squares and maximum likelihood estimates in the exponential family. J Am Stat Assoc 71(353):169–171

Edwards AWF (1972) Likelihood. Cambridge University Press, Cambridge

Edwards AWF (1974) The history of likelihood. Int Stat Rev 42(1):9–15

Geyer (2003) Stat 5102 notes: Fisher information and confidence intervals using maximum likelihood. www.stat.umn.edu/geyer/old03/5102/notes/fish.pdf. Downloaded 4/13/2014

Guo G (1996) Negative multinomial regression models for clustered event counts. Sociol Methodol 26:113–132

Hauer E (2001) Overdispersion in modeling crashes on road sections and in empirical Bayes estimation. Accid Anal Prev 33:799–808

Hyndman RJ, Koehler AB (2006) Another look at measures of forecast accuracy. Int J Forecast 22:679–688

Schermelleh-Engel K, Moosbrugger H, Müller H (2003) Evaluating the fit of structural equation models: tests of significance and descriptive goodness-of-fit measures. Methods Psychol Res Online 8(2):23–74

Stigler SM (1986) The history of statistics. Harvard University Press, Cambridge

Adding Variables

9

Abstract
In this chapter the question is whether to add a variable to the model equation and how to do so. The key concept introduced is that of bias-in-use; it is that bias which arises when the user has information about the value of a safety-related variable which is not in the model equation. The purpose of adding a variable to the model equation is to reduce the bias-in-use. To determine whether adding a variable would reduce the bias-in-use a Variable Introduction Exploratory Data Analysis is conducted. The addition of a variable to the model equation requires a modification of the C-F spreadsheet following which all parameters are reestimated. This is done for the variables AADT, Terrain, and Year. The use of the Negative Multinomial likelihood function with panel data is demonstrated.

9.1 When to Add a Variable

So far Segment Length was the only variable in the model equation. Predictably, the resulting fit was bad. That the number of accidents depends on the amount of traffic is intuitively obvious and empirically substantiated. It is therefore clear that a traffic variable should be in the model equation. Intuition, however, is insufficient when the question is asked about other variables. Should all variables that are thought to affect accident occurrence be in the model equation? Should all variables for which there are data be included? Should one add only those variables the parameters of which are statistically significant or only those that materially improve the fit?

E. Hauer, *The Art of Regression Modeling in Road Safety*,
DOI 10.1007/978-3-319-12529-9_9

The variable inclusion–exclusion decision is a part of the model specification process[1] and what is needed is some clear reasoning and guidance.

In Sect. 1.5 the distinction was made between the "research" and the "applications" purposes of SPF development. The choice of modeling purpose was shown earlier to affect the modeler's attitude to the choice of the objective function. The two perspectives also differ in the considerations that influence the decision of whether to add a variable to the model equation. From the "research" vantage point the purpose of a regression model is to predict by how much a manipulation of a predictor variable will change the dependent variable. This change depends on the size of the parameters. This is why, from the research perspective, focus is on the quality of parameter estimates. With this focus in mind, variables the parameters of which cannot be shown to be statistically different from zero are often not added to the model equation. Alternatively, variables of the least statistically significant parameters are often deleted from it. The "research" perspective offers no unique statistical procedure for adding or deleting variables from a model equation and what procedures are used rely on "a great deal of personal judgment" (Draper and Smith 1981, p. 294). In some cases variables the parameters of which seem to be incorrect are summarily deleted from the model equation.[2]

From the "applications" perspective it is not parameter accuracy, accord with intuition, or statistical significance that matter. Now the purpose of adding a variable to the model equation is to increase the accuracy with which $E\{\mu\}$ is estimated or predicted and to reduce the magnitude of the $\sigma\{\mu\}$. With this purpose in mind the "to add or not to add" question is viewed in the light of necessary and sufficient conditions. Two "necessary" and one "sufficient condition" will be examined.

9.1.1 The Necessary Conditions

For the practical tasks of road safety management it is important that the estimate of $E\{\mu\}$ be unbiased. One kind of bias, the "bias-in-fit," was discussed earlier in Sect. 7.3. That bias arises when in some region of a predictor variable the fitted values are systematically larger or smaller than the observed values. A different kind of bias is the "bias-in-use." This bias arises when a safety-related variable about which the

---

[1] Narrowly construed model specification in regression "…refers to the determination of which independent variables should be included in or excluded from a regression equation" (Allen 1997, p. 166). More broadly the process of model specification consists of selecting an appropriate functional form and of choosing the predictor variables. Common errors of model specification are (1) choosing an incorrect functional form, (2) omitting predictor variables which have a relationship with the dependent variable, (3) including irrelevant predictor variables, and (4) assuming that the predictor variables are measured without error. If an estimated model is misspecified, it will produce biased and inconsistent estimates. (An estimator is consistent if as the number of observations increases the estimates get closer and closer to the true value).

[2] See e.g., Gross et al. (2013, p. 236).

user has information is absent from the model equation or when in the model equation there is a variable for which the user has no information.[3]

To illustrate suppose that we are interested in the safety during 1994–1998 of a certain 0.5 miles long road segment of a rural two-lane road in Colorado with AADT = 5,000. Using the parameter estimates from Fig. 8.6, $\hat{E}\left\{\mu \middle| L = 0.5 \text{ miles}\right\} = 1.656 \times 0.5^{0.871} = 0.905$ I&F accidents in 1994–1998. This would be an unbiased estimate of the μ of this segment if only its length (L) was known. But for the road segment of interest its AADT is also known. When AADT will be added to the model equation in Sect. 9.3 it will turn out that $\hat{E}\{\mu|L = 0.5$ miles, AADT = 5,000$\} = 1.724$ I&F accidents in 1994–1998. Were one to estimate the $E\{\mu\}$ using the model equation without AADT the bias-in-use would be the difference $1.724 - 0.905 = 0.819$ I&F accidents in 5 years.[4]

Another way to describe magnitude of the bias-in-use is by the "(With AADT)/(Without AADT)" ratio in Fig. 9.1. At point A the ratio of $\hat{E}\{\mu\}$s is $1.724/0.905 = 1.9$ and not having AADT in the model equation would cause considerable bias-in-use. However, not having AADT in the model equation would not matter at point B where AADT is about 2,500.

The message of Fig. 9.1 is that, whatever the Segment Length, the "Without AADT" SPF is only good in a narrow range of AADTs and is bad everywhere else. Inasmuch as the SPF is to be used for application with various AADTs, it is clear that estimates of $E\{\mu\}$ produced by a model equation without AADT have a large bias-in-use and are therefore of no practical use.

The same, of course, can be said for any variable the absence of which from the model equation would, in some region, have "With/Without" ratios that are much different from 1.[5] What matters in applications is not the statistical significance of a parameter but whether the absence of a variable from the model equation would cause practically important bias-in-use.

The obverse is also important. To compute the value of the dependent variable on the left-hand side of the model equation one must have values for all variables on its right-hand side. Suppose, for example, the modeler included in the model

[3] The Bias-In-Use is not what in statistics is called the "Omitted-Variable Bias." The OVB is a bias in the parameters which occurs when a model incorrectly leaves out one or more important causal variables. The bias-in-use pertains to the estimate of $E\{\mu\}$ and arises when the variables in the SPF do not match the known safety-related traits of the unit or the population the $E\{\mu\}$ of which is of interest. From the "research" perspective it is the OVB that matters; from the "applications" perspective it is the bias-in-use that is important.

[4] The following may serve as a succinct definition of bias-in-use: By "obvious observation 1" in Sect. 3.4 if A and B are variables and B is safety-related then $E\{\mu|A \text{ and } B\} \neq E\{\mu|A\}$. The "bias-in-use" is the difference $E\{\mu|A \text{ and } B\} - E\{\mu|A\}$.

[5] Situations in which variables absent from the model equation give cause for concern are commonly encountered and familiar to practitioners. To illustrate, the practitioner will sense that something is amiss when, say, the road segment of interest is a sag curve preceded by a long steep grade but grade and vertical curvature variables are not in the SPF.

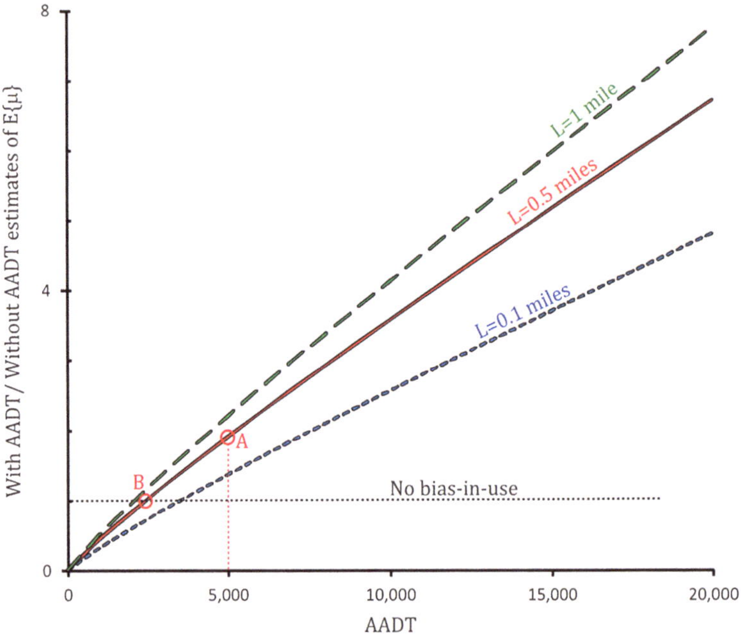

Fig. 9.1 The "With/Without" ratios for AADT

equation a variable called "Roadside Hazard Rating (RHR)[6]." Information about RHR is seldom found in databases. To use such an SPF the practitioner must either inspect the site of interest and determine its RHR value or plug into the model equation an educated guess. When a guess is used the resulting $\hat{E}\{\mu\}$ will be biased, unless the guess is a lucky one.

All this was noted in Sect. 3.4 under "obvious observation 2": "*For the* $\hat{E}\{\mu\}$ *of a population to be an unbiased estimator of the* μ *of a unit of interest, the known traits of that unit must be the same as the traits which define the population of the units for which* $E\{\mu\}$ *is the mean.*" This was said to "*lead to a 'not-so-obvious' and yet important conclusion. Namely, that what traits (variables) are to be used in a SPFs depends on the use for which the SPF is intended and on the information the user has.*"

The bias-in-use is 0 if the variables in the SPF are the same as the data that are available for the practical task in which the SPF is used.

There is a close logical link between "bias-in-use" and the concept of a variable being "safety related." A variable was said to be "safety related" if when it changes then the μ of the unit changes.[7] Here the same concept has a more targeted meaning

[6] See e.g., Zegeer et al. (1988) and Harwood et al. (2000), Appendix D.

[7] See Sect. 1.2.

and applies to change in $E\{\mu\}$ due to the addition of a variable. Let A and B be variables and the "bias-in-use" be the difference $E\{\mu|A \text{ and } B\} - E\{\mu|A\}$. This difference is 0 when $E\{\mu|A\} = E\{\mu|A \text{ and } B\}$ for all values of B. That is, when the bias-in-use is 0 for all value of B of practical interest, then variable B is not safety related. Conversely, when the absence of variable B from the model causes bias-in-use, then $E\{\mu|A\} \neq E\{\mu|A \text{ and } B\}$ and therefore B is safety related.

Whether a variable is safety related can be examined empirically. Thus, for example, if the residuals of the "without AADT" model tend to be positive for segments with large AADT, negative for segments with small AADT, and near zero in-between, then AADT is safety related. Because information about AADT is usually available in practice, its absence from the model equation will cause bias-in-use.

It follows that the necessary conditions for a variable to be introduced into the model equation are two:

(i) The variable must be safety related
(ii) Information about the variable is available to users of SPFs.

Condition (ii) has consequences for report writing. The process of SPF development was described in Sect. 6.1 as a gradual build up of the model equation to which traits (variables) are added one after another. This will be the pattern followed in this chapter; first AADT will be added to L, then "Terrain," and later "Year." Suppose that variable "*A*" was introduced into the model equation first, followed by "*B*", "*C*", and then "*D*." Thus, in the course of SPF development the modeler would have obtained four models; one with *A*, the next with *A*&*B*, etc. Had the modeler reported all four model equations, the user could choose that model of the four for which he or she has data. Unfortunately, modelers usually report only the final and "fully loaded" model. This is wasteful. While the user may not have information about all four variables in the final model and therefore cannot use it, the interim models with *A*, *A*&*B*, or *A*&*B*&*C* might have been of use.

The modeler cannot know in what circumstances the SPF will be used and what information future users will have. The least the modeler can do is to report all model equations and parameter estimates as variables are introduced, one after another. The best that could be done would be to identify commonly available variable bundles and for these to estimate the corresponding models.

9.1.2 The Sufficient Condition

Should all variables that meet the necessary conditions be added to the SPF? A reasoned answer depends on the balance between what is gained and what is lost by doing so. The basic consideration is governed by (9.1).

$$\text{Expected Mean-Square Error of } \hat{E}\{\mu\} \equiv (\text{bias-in-use})^2 + V\{\hat{E}\{\mu\}\} \qquad (9.1)$$

The gain is that of reducing the bias-in-use; the loss is in the possible increase of the variance of the estimate of $E\{\mu\}$.

Why adding a variable increases the $V\{\hat{E}\{\mu\}\}$ was easy to explain when, in Chap. 3, $\hat{E}\{\mu\}$ was the sum of accidents in a bin divided by the number of units in that bin. Adding a variable meant the partitioning of the bins and thereby reducing the number of units used to estimate the $E\{\mu\}$'s.[8] However, when estimation of $E\{\mu\}$ is by a parametric model equation, when the populations to which the $E\{\mu\}$'s pertain are imagined, the explanation is different. Each predictor variable the value of which is uncertain (e.g., the AADT) and each added parameter estimate have a variance. The more numerous are the predictor variables and the parameters in a model equation, the larger will the $V\{\hat{E}\{\mu\}\}$ tend to be. However, because interactions are complex and correlations many, the net effect is difficult to foresee.[9] More on this is in Chap. 11.

Balancing gain and loss is not easy for two reasons. The first difficulty is that the balance depends on circumstances and on judgment. To explain, consider the vicinity of point *B* in Fig. 9.1 where the bias-in-use is small and there is little to be gained by adding AADT to the model equation. Inasmuch as the addition of AADT might increase the $V\{\hat{E}\{\mu\}\}$ there will be a region near point *B* where the loss due to the addition of AADT likely outweighs the gain. In contrast, where the bias-in-use is large, e.g., to the right of point *A* in Fig. 9.1, the gain of adding the AADT is almost certainly larger than the loss. Thus, when AADT is about 2,500 a model without AADT might estimate better while the opposite is true when AADT is larger than, say, 5,000. Since adding AADT to the model equation will degrade estimates of $E\{\mu\}$ in some range of AADT and improve them elsewhere, the "to add or not to add" question is difficult to answer without the exercise of judgment.[10]

---

[8] Using the numerical results of the EDA as illustration it was easy to state as "Obvious Observation 3" that "The larger is the number of traits that define a real population, the fewer are the observations from which its $E\{\mu\}$ is estimated and the larger tends to be the standard error of the estimate of $E\{\mu\}$."

[9] Consider a multiplicative model equation in which expression *A* accounts for the contribution of variables $X_1, X_2, \ldots, X_n$ and expression *B* accounts for the contribution of a new variable X_{n+1}. Because of the uncertainties in the *X*'s and their parameters, both *A* and *B* are random variables. Let $\sigma\{A\}$ and $\sigma\{B\}$ be their standard deviations and $\rho\{A,B\}$ their correlation coefficient. With this, $\left(\frac{\sigma\{AB\}}{AB}\right)^2 = \left(\frac{\sigma\{A\}}{A}\right)^2 + \left(\frac{\sigma\{B\}}{B}\right)^2 + 2\frac{\sigma\{A\}\sigma\{B\}}{AB}\rho\{A,B\}$. If *A* and *B* are uncorrelated then $(\sigma\{AB\})^2 = B^2(\sigma\{A\})^2 + A^2(\sigma\{B\})^2$, If expression *B* which accounts for the effect of the added uncorrelated variable has an average value of 1, the consequence of adding it to the model is to increase the variance by $A^2(\sigma\{B\})^2$, While it is tempting to assume that *A* and *B* are uncorrelated, this is seldom so. The addition of *B* will always change the scale parameter which is part of *A*. It is therefore difficult to foresee how the addition of a variable will affect the $V\{\hat{E}\{\mu\}\}$.

[10] Lehmann (1990, p. 162) suggests using the average squared prediction error saying that: "The best fitting model is of course always the one that includes the largest number of variables. However, this difficulty can be overcome …, for example by selecting the dimension k which minimizes … the expected squared difference between the next observation and its best prediction from the model. This measure has two components: *E*(squared prediction error) = (squared bias) + (variance). As the dimension *k* of the model increases, the bias will decrease. At the same time the variance will tend to increase since the need to estimate a larger number of parameters will increase the variability of the estimator."

Another approach might be to equate the gain of adding a variable to the reduction in $V\{\hat{\mu}_i\}$. The rationale for doing so is the same as that which led to (2.2) in Sect. 2.3. Namely, that if the $\hat{E}\{\mu\}$ which the SPF provides is to serve as the estimate of the μ of a certain site i ($\hat{\mu}_i$) then

$$V\{\hat{\mu}_i\} = V\{\hat{E}\{\mu\}\} + V\{\mu\} \tag{9.2}$$

Thus, if adding the variable to the model equation increases the $V\{\hat{E}\{\mu\}\}$ by less than it diminishes the $V\{\mu\}$ then, for site i, the necessary condition is considered to be met. If the condition is met for most sites for which the $\hat{E}\{\mu\}$s will help with the estimation of μ then the necessary condition is met.

The second difficulty is that to use (9.1) and (9.2) one has to have an estimate of $V\{\hat{E}\{\mu\}\}$ and obtaining it is not simple. As is already clear and will become increasingly more evident, what $E\{\mu\}$ is estimated to be depends on an assortment of assumptions. One can estimate the $V\{\hat{E}\{\mu\}\}$ as if these assumptions were true. However, inasmuch as different assumptions, all plausible, lead to different estimates of $E\{\mu\}$, the assumption-based estimate of $V\{\hat{E}\{\mu\}\}$ is certain to be too small. It is best to postpone discussion about $V\{\hat{E}\{\mu\}\}$ till after the nature and role of most assumptions has been more fully described. Therefore, how $V\{\hat{E}\{\mu\}\}$ can be estimated and by how much it is increased when a variable is added to the model equation will be discussed only in Chap. 11. As a consequence, in what follows, the variable addition decision will be based on the necessary conditions only; that is, on whether the variable is safety related and on whether information about it is available to users.

If a variable is safety related and is the source of bias-in-use, adding it to the model equation will reduce the size of residuals, will improve the fit, and will change the $\hat{E}\{\mu\}$. If a variable is not safety related and therefore is not a source of bias-in-use, it should not be in the model equation. The determination of whether a variable is safety related and what is the form of its relationship with the $E\{\mu\}$ is the task of the "Variable Introduction Exploratory Data Analysis" discussed next.

9.2 The Variable Introduction EDA: Is AADT Safety Related?

The necessary condition for adding a variable to the model equation is that it be safety related. To determine the safety relatedness of some variable one has to do a Variable Introduction Exploratory Data Analysis – VIEDA for short. The purposes and tasks of a VIEDA are two:

(i) To determine whether the residuals indicate that the variable is safety related and, if yes
(ii) What function could represent the new variable in the model equation

Here the specific question is whether to add the Average AADT to the model equation. The data for this VIEDA come from the C-F spreadsheet in Fig. 8.6 to which, in Fig. 9.2, the Average AADT was added as column C. One way of going about task (i) is to create a pivot table[11] with the sums of observed and fitted values (columns I and J) for some AADT bins.

It is apparent that for segments with AADT < 2,000 the model consistently predicts more accidents than what was observed and that the opposite is true for larger AADTs. The graph of the "Observed/Fitted" ratio as a function of AADT is in Fig. 9.3. Not only is AADT seen to be safety related, the relationship is also quite orderly.

The choice of the "Observed/Fitted" ratio for examining the orderliness of the relationship serves a purpose. This is the ratio that brings the fitted values into line with what was observed. Thus, for example, the 1,898.7 fitted accidents in the 0 < AADT < 500 bin of Fig. 9.2 need to be multiplied by "Observed/Fitted" ratio of 0.25 to make them into 473 observed accidents. Therefore, if AADT is to be

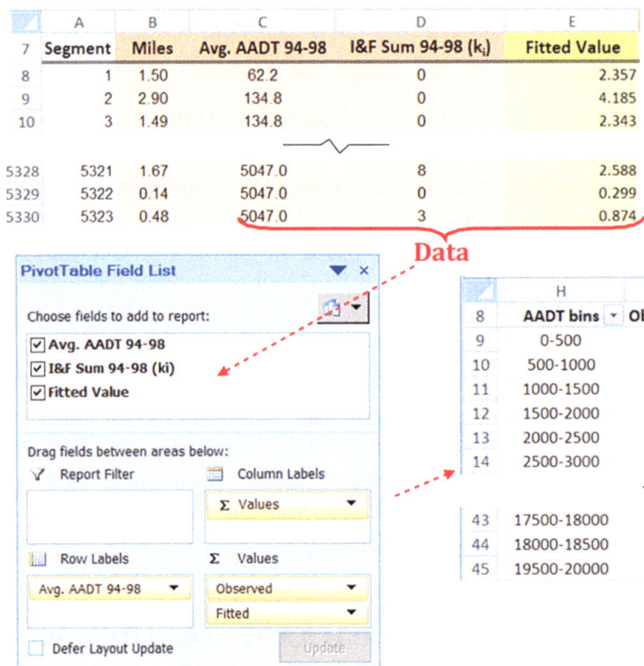

Fig. 9.2 Pivot table for VIEDA about AADT

[11] How to work with Pivot Tables was explained in Sect. 3.3. To download this spreadsheet go to http://extras.springer.com/ and enter the ISBN of this book. Look in the "Spreadsheets" folder for "Chapter 9. VIEDA for AADT.xls or xlsm."

introduced into the model equation by a multiplicative function, it should resemble relationship in Fig. 9.3.[12]

An alternative way to check whether the relationship is orderly is to forgo the grouping into bins and smooth the ungrouped data by, say, the Nadaraya–Watson procedure instead.[13] The result is in Fig. 9.4. Up to about AADT = 10,000 the

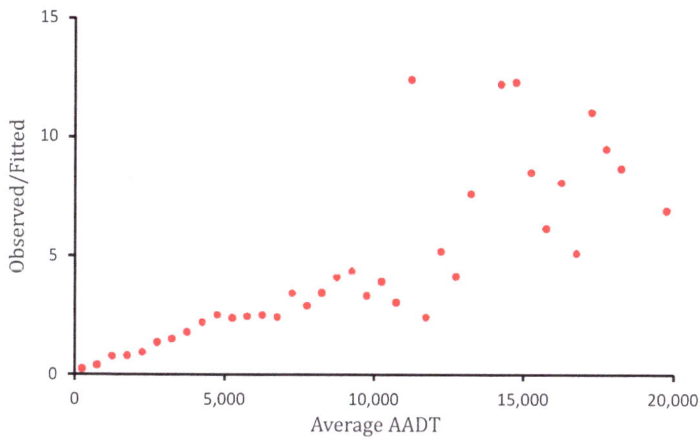

Fig. 9.3 The relationship between Average AADT and the Observed/Fitted ratio

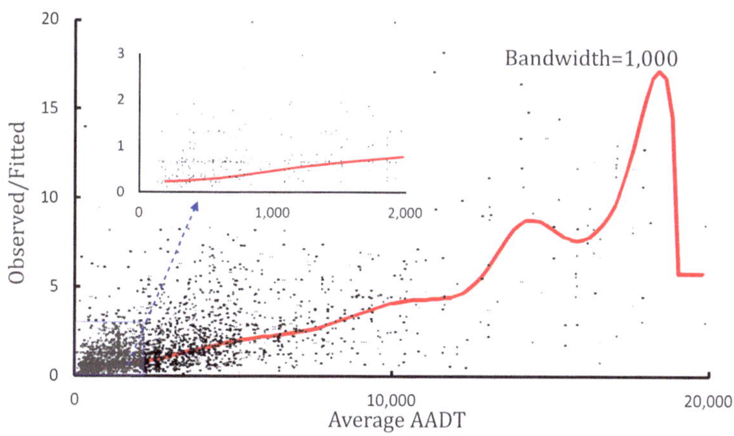

Fig. 9.4 Smoothed relationship between AADT and Observed/Fitted ratio

---

[12] A new variable is often introduced into the model equation by a function that multiplies the previous expression. In this case the Observed/Fitted ratio is of interest. When the new variable is introduced into the model equation by an additive function "case b" in Hauer (2004) should be used.

[13] See Sect. 4.3.

existence of an orderly relationship is evident. The undulations for larger AADTs reflect the sparsity of the data rather than the absence of an orderly relationship.

Purpose (ii) of VIEDA is to suggest the function by which the variable should be represented in the model equation. The points in Fig. 9.3 and the curve in Fig. 9.4 suggest a straight line or an upward bending curve. The insert in Fig. 9.4 which depicts the near-origin detail hints at a positive intercept. The function AADT^{β_2} is a reasonable initial guess.

In sum, since information about the AADT is usually available the two necessary conditions for adding Average AADT to the model equation are met. The next task is to show how to add a function of Average AADT to the model equation and to the C-F spreadsheet.

9.3 How to Add a Variable to the C-F Spreadsheet

Adding a variable to the C-F spreadsheet is straightforward. For didactic reasons[14] the influence of the Average AADT will be initially represented by the "power function": AADT^{β_2}. Thus, the model equation to be fitted is $\hat{E}\{\mu\} = \beta_0 L^{\beta_1}\text{AADT}^{\beta_2}$. The task is to incorporate the power function for Average AADT into the C-F spreadsheet of the NB fit from Fig. 8.6 and then to reestimate all parameters. How this is done is shown in Fig. 9.5.[15]

The top part is the earlier NB fit with only Segment Length in the model equation. In the bottom part the AADT data were added as column C, the new parameter β_2 is in cell G2, and the fitted values in Column G are the product of the scale parameter in E2 and the factors L^{β_1} and AADT^{β_2} in columns E and F.[16] With the newly estimated parameters[17] the SPF for injury and fatal accidents in 5 years is:

$$\begin{aligned} E\{\mu\} &= 0.00158 L^{1.078}\ \text{AADT}^{0.909} \\ \hat{V}\{\mu\} &= \frac{\left(\hat{E}\{\mu\}\right)^2}{2.202L} \end{aligned} \tag{9.3}$$

[14] The didactic reasons are two. First, that the power function X^{β} is widely used and, as such, can be a benchmark against which to judge alternative functions to be examined later. Second, since the power function starts at the origin but the VIEDA suggests the presence of an intercept, the use of the power function might be reflected in the CURE plot and thereby illustrate its diagnostic use.

[15] To download this spreadsheet go to http://extras.springer.com/ and enter the ISBN of this book. Look in the "Spreadsheets" folder for Chapter 9. NB fit for L and AADT and CURE plots.xls or xlsx

[16] Keeping the factors in separate columns makes experimentation with alternative functional forms easier.

[17] When running Solver the parameter estimates from the top part of Fig. 9.5 were used as initial guesses except that 0.001 was used for β_0. Because now the parameters differ by several orders of magnitude, it is important to activate the "automatic scaling" option in Solver. See discussion in Sect. 5.4.

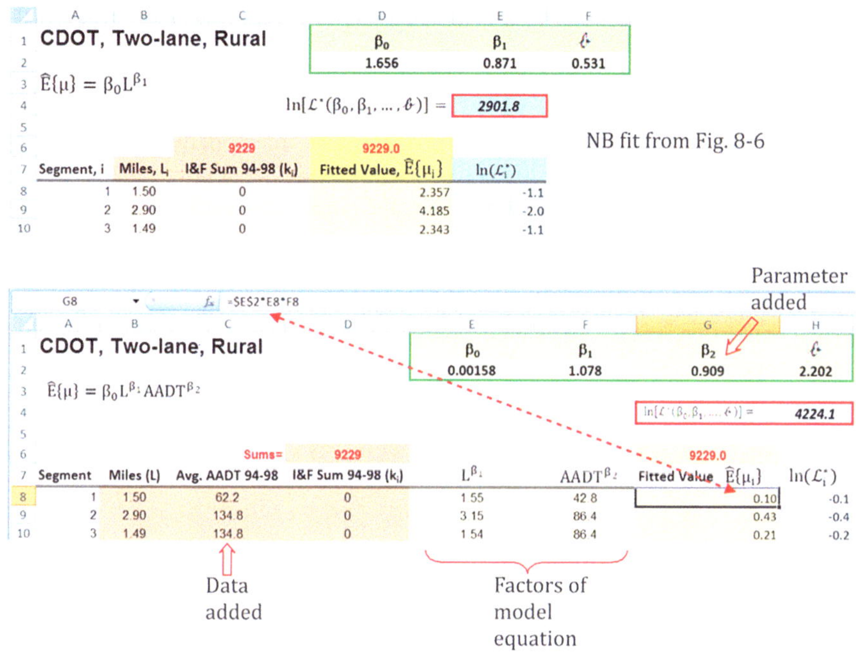

Fig. 9.5 Adding AADT to the model equation and the C-F spreadsheet

9.4 The Omitted Variable Bias

One consequence of adding AADT to the model equation was to change the estimate the parameter β_1 from 0.871 to 1.078.[18] This is way out of the 95 % confidence interval surrounding the earlier estimate[19] and underscores a general truth which modelers recognize from experience. Namely, that when a new variable is added to the model equation, previously estimated parameter values tend to

[18] It was said in Sect. 6.7.1 that: "... should at a later stage in modeling the estimate of β_1 become indistinguishable from 1 that would be a sign that L reflects only the influence of segment length and that the influence of other safety-related variables has been satisfactorily accounted for." Now, after the introduction of AADT, it seems that β_1 may indeed be fluctuating around 1. Should it remain so after more variables are added, it may be then be best to set the exponent of L to 1.

[19] In Sect. 6.6.1 the 95 % confidence interval was 0.82–0.91.

change. This "Omitted Variable Bias"[20] (OVB) is one of the main sources of the "modeling uncertainty."[21]

The larger the correlation of the newly introduced variable with those previously used,[22] the more difficult it is to tell apart their influence on the $E\{\mu\}$.[23] If so, whatever are the parameter estimates at some stage of modeling, they must be viewed as provisional; their value is likely to change when a new variable will be added to the model equation. By how much the parameter estimates would change cannot be known beforehand. If the change is substantial it is a sign that the model is still subject to the "Omitted Variable Bias." If so, till the OVB issue is satisfactorily addressed, one should place no trust in the commonly reported standard errors, t-statistics, and p-values which purport to describe that accuracy of parameter estimates.

One can begin to trust the parameter values if they do not change much when new variables are introduced into the model equation.

The habit of reporting parameter accuracy as if OVB and other sources of modeling inaccuracy did not exist is intellectually wanting. Want turns into harm when untrustworthy parameter estimates are used to predict how safety would change if some variable value was changed.[24] Thus, for example, in the horizontal curve design example of Sect. 6.7.1 the exponent of L was 0.86 and the upper limit of its 95 % confidence interval was well below 1. However now that AADT was added, the exponent of L seems to be 1.08. If that was the final estimate the considerations in the design example would be reversed. Alas, the estimate is not final. After "Terrain" will be added the estimate of β_1 will drop back to 0.985 and after "Time" will be accounted for it will be further reduced to 0.969.

---

[20] The omitted-variable bias occurs when a model leaves out one or more safety-relevant variables. The bias pertains to estimates of parameters. When statistical packages report the standard error of parameter estimates they do so under the grossly unrealistic assumption that all causal variables are in the model equation and that the model equation is an exact representation of the relationship underlying the data. For these reasons the usually reported estimates of parameter accuracy cannot be trusted.

[21] The other source discussed earlier was the choice of objective function (see Table 8.1). For a discussion of "modeling uncertainty" see Sect. 6.6.2.

[22] In road safety, correlation between traits (variables) is a pervasive part of reality. One only has to think about the role played by traffic volume in determining the traits of roads and intersections. In our data the correlation between Segment Length and AADT is −0.17; Segments with more traffic tend to be shorter. Even this modest correlation is sufficient to cause a sizeable modeling uncertainty about the parameter of L.

[23] In curve-fitting this phenomenon goes under the name collinearity. When variables are highly correlated the regression parameters change erratically in response to small changes in the model or the data.

[24] Why such use of SPFs is problematic was discussed at length in Sect. 6.7.

The importance of the realization that the OVB is omnipresent and its effect can be sizable is threefold. First, when SPFs are used for the practical purposes described in Sects. 1.3 and 1.4, and if only the statistical accuracy of parameter estimates is recognized when $\sigma\{\hat{E}\{\mu\}\}$ is estimated,[25] it will be an underestimate. Second, when parameter estimates are used as the source of Crash Modification Factors,[26] the users can have no clear idea about how inaccurate these numbers really are. Third, in a more constructive vein, when the parameters in the model equation attain a measure of stability and cease to change significantly as new variables are added, confidence in their value can grow.

A comment about the evolution of the estimates of β_1 is in order here. The natural starting assumption might be that the expected number of crashes is proportional to Segment Length. This would be true were it not for the fact that the length of segments in databases depends on safety-related factors such as intersection density, changes in number of lanes, and differences in speed limit. Indeed, at the conclusion of the initial EDA the impression was of a nonlinear perhaps upward-bending relationship between crash frequency and segment length. Attempting to let the data to speak for themselves, the power function L^{β_1} was chosen to represent the relationship. With this as the model equation and with Segment Length as the only variable the data favored a downward bending function. Indeed, all estimates of β_1 in Table 8.1 are all smaller than 1. However, after AADT was added to the model equation the estimate of β_1 turned out to be larger than 1 and the curve seemed to be upward bending. After Terrain will be added to L and AADT the estimate of β_1 will change to 0.985 bringing us back full circle, close to the original proportionality assumption. This lends credence to the opinion that SPF modeling is best viewed as a process during which the model gradually takes shape; that conclusions are always provisional and could change if some of the choices which the modeler made were to be altered and if information about additional safety-related traits was available.

The addition of AADT to the model also changed the estimate of $\mathscr{b}$ from 0.531 to 2.201. The larger is $\mathscr{b}$ the smaller is the variance of the μ's.[27] That is, the μ's in a population of units defined by both Segment Length and AADT have a much smaller variance than those of a population of units defined by Segment Length only. The smaller is $V\{\mu\}$ the more influential is the SPF estimate in all applications. Therefore, changes to the model equation that make $\mathscr{b}$ larger are good for practice.

The hope was that adding AADT to the model equation will improve on the heretofore unsatisfactory fit; whether it did is examined next.

---

[25] See, e.g., Wood (2005), Lord (2008) and Lord et al. (2010).

[26] In spite of the reservations discussed in Sect. 6.7.

[27] Recall that $V\{\mu\} = (E\{\mu\})^2/(\mathscr{b}L)$.

9.5 A Few CURE Plots

To see the extent to which adding a variable to the model equation improved the SPF and what problems of fit remain, the CURE plots for every variable have to be checked.[28]

The comparison of "without AADT" and "with AADT" plots for *L* is in Fig. 9.6. As is evident, the addition of AADT was a significant improvement. The nearly horizontal stretches of the new CURE plot are regions of small bias-in-fit and they cover most segment lengths.

The two remaining regions of concern are marked by ellipses. Ellipse *A* is where the observed accident counts tend to be larger than the fitted values. It is possible that this can be remedied by choosing a different functional form to account for the contribution of *L*.[29]

Ellipse *B* is over what seems to be a nearly vertical drop which could be concealing outliers. An outlier is a road segment for which the accident count is very different from the fitted value. In the CURE plot it would show as a large vertical gap. Figure 9.7 examines the suspect region of ellipse *B* in detail.

There is no outlier here. What in Fig. 9.6 appeared to be a solid vertical drop is in fact an agglomeration of many small ones. There are many road segments which are 0.89, 0.90, and 0.91 miles long and on which the accident count tends to be consistently smaller than are the fitted values. What makes segments of these lengths peculiar in this way is unclear. It is not likely that the peculiarity will diminish when alternative functional forms are tried; perhaps it will shrink when additional variables will be introduced.

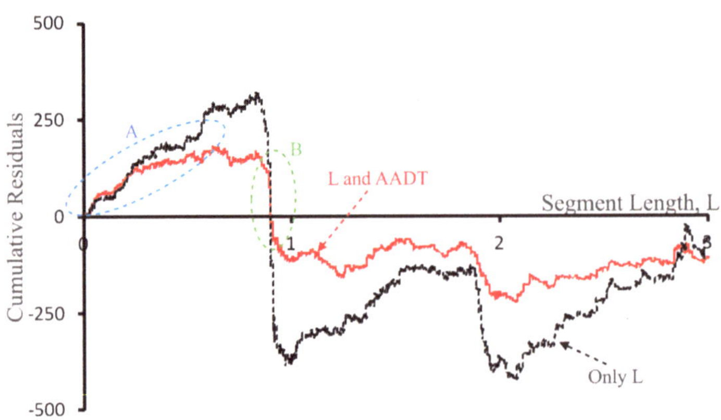

Fig. 9.6 Comparing two CURE plots for *L*

[28] To download this spreadsheet go to http://extras.springer.com/ and enter the ISBN of this book. Look in the "Spreadsheets" folder for "Chapter 9. NB fit for L and AADT and CURE plots.xls or xlsx."

[29] The examination of alternative functional forms will be the subject of Chap. 10.

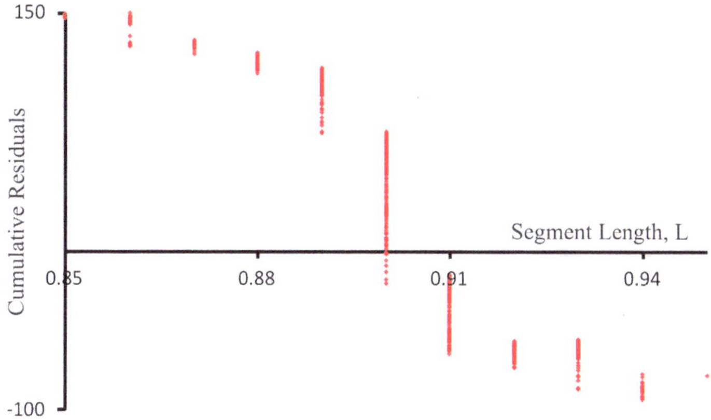

Fig. 9.7 Is there an outlier under ellipse *B*?

The other CURE plot that has to be examined, that with respect to the AADT, is in Fig. 9.8.

The nearly horizontal stretch from *C* to *D* looks like a benign random walk. Between the origin and point *A* there are more accidents than what is predicted. This excess might disappear once other variables are added.[30] Alternatively, it could be corrected by using a positive intercept as was intimated earlier in Fig. 9.4. The main problem is the ill fitting short stretch between *B* and *C* (1,550 < AADT < 2,600) where the model overpredicts by about 26 %. To correct this bias one would have to find a function which predicts more accidents then the Power function for AADT < 1,500, fewer in the 1,500 < AADT < 2,600 range, and about the same for AADT > 2,600. A simple function meeting these requirements may be difficult to find.

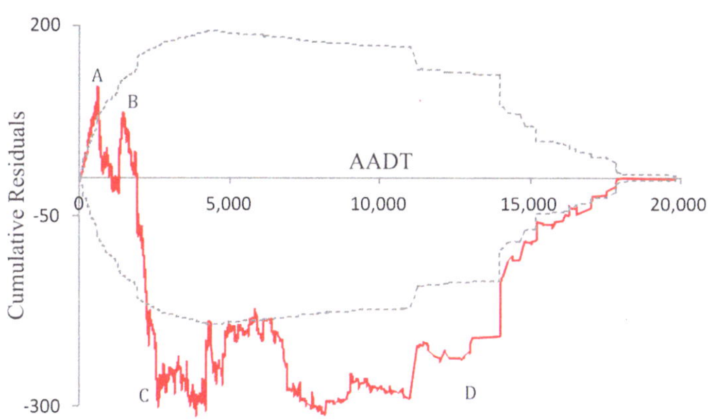

Fig. 9.8 The CURE plot for AADT

[30] For example, if these low-AADT segments are mostly in mountainous terrain where, with the same *L* and AADT more accidents are predicted than in, say, flat terrain.

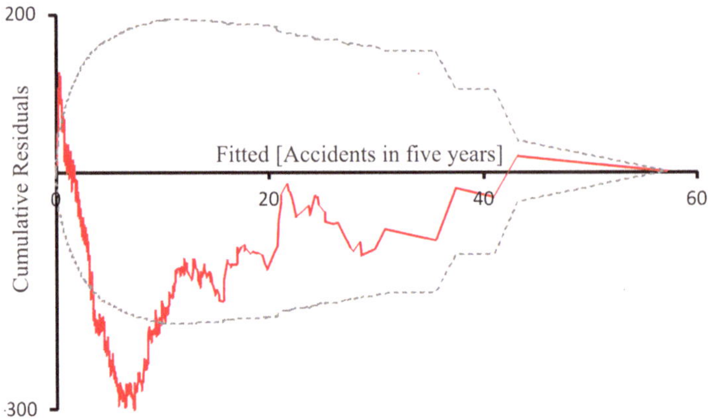

Fig. 9.9 The CURE plot for "Fitted Values"

The CURE plots for *L* and AADT contain hints about how to improve the representation of these variables in the model equation. However, the overall fit of the SPF is best judged by the CURE plot for fitted values.[31] This "fitted-values" CURE plot is in Fig. 9.9; it too is unsatisfactory.

9.6 Adding Variables: Terrain

The Colorado data contain information about whether a segment is in Flat, Rolling, or Mountainous (F, R, or M) terrain.[32] The Terrain variable is associated with traits such as grade, curvature, and roadside, which are believed to affect crash frequency and severity. Therefore one may expect "Terrain" to be a safety-related proxy variable. Whether Terrain is in fact safety related, i.e., whether its absence from the model equation is a source of bias-in-use, is easily checked. Use of the fitted values from Fig. 9.5 and of the pivot table magic leads to Table 9.1.[33]

Table 9.1 VIEDA for "Terrain"

Terrain	Observed accidents	Fitted values	Observed/Fitted
Flat	831	1,424.9	0.58
Mountainous	4,806	3,682.6	1.31
Rolling	3,592	4,121.5	0.87

[31] In terms of computation, all that is required is to sort by the "fitted value" (see, e.g., column D in Fig. 7.3) instead of sorting by a variable value (column B in the same figure).

[32] See Fig. 3.1.

[33] To download this spreadsheet go to http://extras.springer.com/ and enter the ISBN of this book. Look in the "Spreadsheets" folder for "Chapter 9. Terrain Pivot.xls or xlsx."

The message is clear: In flat terrain there are fewer accidents than what (9.3) predicts and using it to estimate the $E\{\mu\}$ of a segment in flat terrain would be usually an overestimate. Similarly, for a segment in mountainous terrain (9.3) would under-predict. Were "Terrain" not in the model equation there would be considerable bias-in-use. Inasmuch as information about the terrain in which a road segment is situated is either known or easily ascertained, the necessary conditions for the inclusion of Terrain into the model equation are fulfilled.

The "Terrain" variable can be introduced into the model equation in several ways only one of which will be examined here.[34] A simple way of accounting for its influence is by a multiplicative parameter β_{Terrain} in $\hat{E}\{\mu\} = \beta_0 L^{\beta_1}\,\text{AADT}^{\beta_2}\beta_{\text{Terrain}}$. The $\beta_{\text{Terrain}=\text{F}}$ will be set to 1 and the values of β_{R} and β_{M} will be estimated using the Solver. When choosing this way of representing the influence of the Terrain variable the modeler is assuming that the influence of L and AADT can be represented by the same function and the same parameters β_1, β_2 in all terrains and that the overdispersion parameter ℓ is also the same in all three terrains.[35]

To estimate the parameters of the enriched-by-terrain model the earlier C-F spreadsheet in Fig. 9.5 had to be modified in several respects as is shown in Fig. 9.10[36]: the terrain data were added as column D; two new terrain parameters were placed into cells G2:H2; a corresponding multiplier factor was added as column H; and this factor was added to the other two factors (one for L and the other for AADT) to compute the fitted value in column I. With these modifications the Solver produced new parameter estimates.

The two "Terrain" multipliers are rather large.[37] Now the model equation is:

$$\hat{E}\{\mu\} = 0.00160L^{0.985}\,\text{AADT}^{0.835}(1 \text{ if } F, 1.642 \text{ if } R \text{ and } 2.495 \text{ if } M)\,\text{acc. in 5 years}$$
$$\hat{V}\{\mu\} = \frac{\left(\hat{E}\{\mu\}\right)^2}{2.823L} \tag{9.4}$$

[34] The question of how the influence of a variable should be represented in the model equation will be discussed in Chap. 10. Thus, e.g., one of the EDA findings in Sect. 3.6 was that the effect of Terrain might depend on both AADT and Segment Length. The question will be how such a dependence can be captured.

[35] As already noted, in the course of modeling the modeler has to make various assumptions. In the course of the continuing illustration it was assumed that the model equation is a product of power functions, that maximizing likelihood is a good objective, that accident counts are Poisson distributed, that the μ's are Gamma distributed, etc. Now another assumption is added: that the influence on $E\{\mu\}$ of L and AADT is the same in every terrain. There is, of course, no prior support for such an assumption. Its validity can be examined by fitting separate models for each terrain.

[36] To download this spreadsheet go to http://extras.springer.com/ and enter the ISBN of this book. Look in the "Spreadsheets" folder for "Chapter 9. NB fit with L, AADT and Terrain as multiplier. xls or xlsx."

[37] A segment in rolling terrain is estimated to have 1.642 times the number of accidents of a segment of the same length and same AADT in flat terrain; the corresponding multiplier for mountainous terrain is 2.495. By the numbers in Table 9.1 the corresponding implied multipliers are 0.87/0.58 = 1.5 and 1.31/0.58 = 2.3.

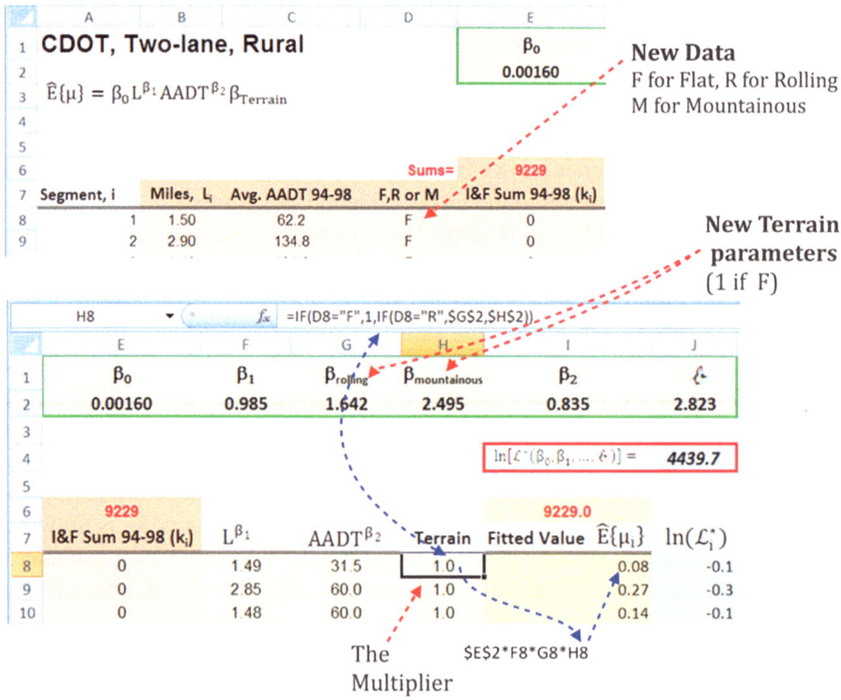

Fig. 9.10 Terrain as a multiplier

The next variable considered for addition to the model equation is "Time." Before this can be done effectively the use of the condensed data for a single 5-year period has to be supplanted by the use of panel data from 13 separate years and the move to the Negative Multinomial (NM) likelihood has to be implemented.

9.7 Panel Data and the NM Likelihood

The original Colorado data contain AADT estimates and accident counts for each of the 13 years 1986–1998.[38] However, to this point, only the "condensed data" were used in which the 1994–1998 were collapsed into a single period such that the average AADT and the sum of accidents for those 5 years as served as data.[39] One reason for using a single period was that the expression for the popular NB likelihood function is derived with one time period in mind.[40] One is assuming

[38] See Sect. 3.2.

[39] See Sect. 3.2 and Fig. 3.2.

[40] See Appendix E.

that within that single period variable values do not change. For this to be plausible the duration of the period is usually kept to just few years; five is already a stretch.

But AADT does change over the years[41] and so do any other traits (average speed, pavement friction, proportion of trucks, average blood alcohol content, seatbelt wearing rate, etc.). Not to use all available data amounts to a loss of information in more ways than one. First, it is obviously better to use more data than less. Second, the multi-year average AADT irons out some useful detail that is present in yearly AADTs. Third, if some safety-relevant trait of a unit changed within the period, if shoulders were paved or the intersection signalized, data for that unit may not be usable in a single period model and yet could be usable in a multi-period model. Fourth, by collapsing yearly data into a single multi-year period the resulting estimates and predictions pertain only to that period as a whole; estimates and predictions for specific years of that period will be biased and lost will be the possibility of extrapolation to years outside that single period.

Cross sectional data about units with variable values available for two or more time periods are called "panel" or "longitudinal" data. When panel data are available and parameters are estimated by maximizing likelihood, the NM[42] likelihood function replaces the NB.[43] The implementation of the NM likelihood function requires changes in the C-F spreadsheet. These are shown in Fig. 9.11.

The Colorado panel data are in the two top tiers. Both Segment Length and AADT were divided by their largest values defining the normalized variables X_1 and X_2 so that all values are between 0 and 1. What previously was the "Average AADT" is now split into 13 columns of X_2 values (in columns C to O). Similarly there are now 13 yearly accident counts (columns P to AB) and, in preparation for use in the log-likelihood formula, given is their sum (in column AC). The three factors of the model equation make up the next tier. Since Segment Length and Terrain do not vary with time these factors have one column each (AE and AS), while the X_2 variable the value of which may change from year to year has 13 yearly factors (columns AF to AR). The product of these factors multiplied by the scale parameter gives the 13 yearly fitted values in the bottom tier.

---

[41] Over the 13-year period 1986–1998 on rural two-lane roads in Colorado the AADT increased, on the average, by about 55 %.

[42] See Sects. 8.3.2, 8.3.3 and Appendices E, G.

[43] With a slight compromise of purity it is possible to use the NB likelihood even with panel data. This approach will be used in the simulations of Chap. 11 where saving on computations matters. To download the spreadsheet where the NM likelihood was used go to http://extras.springer.com/ and enter the ISBN of this book. Look in the "Spreadsheets" folder for "Chapter 9. NM, 13 year panel, *L*, AADT, Terrain, and a common scale parameter.xls or xlsx." To download the spreadsheet where the NB likelihood was used look for "Chapter 9. NB, 13-year panel.xls or xlsx."

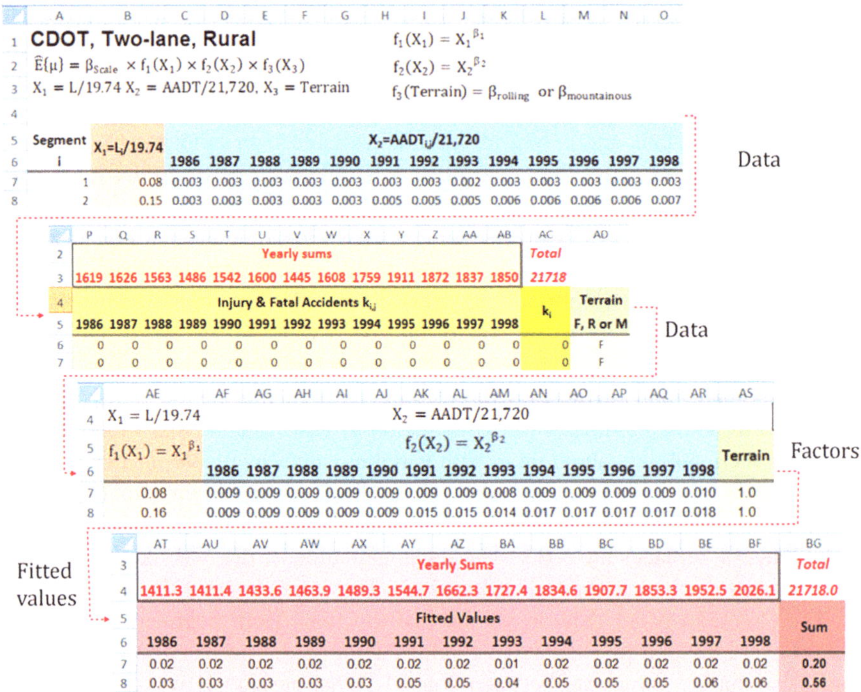

Fig. 9.11 Panel data, factors of model equation, and fitted values

The abridged Multinomial log-likelihood equation from Sect. 8.3.3 is reproduced on the top of Fig. 9.12 and placed in the formula bar as the content of cell BX7. The Solver was used to maximize the content of BX4 and the "By Changing" cells were BS3:BX3. The "Automatic scaling" option was chosen because β_0 differs from the other parameters by several orders of magnitude.[44]

The parameter estimates obtained by maximizing the NB likelihood when the 5 years of condensed data were used (Fig. 9.10) are similar to those obtained using the NM likelihood and 13 years of data (Fig. 9.12).[45] Both sets of parameter estimates lead to very poor CURE plots. Can these improve by choosing a different objective function?

[44] See Sect. 5.4 for detail.

[45] The difference in $\mathscr{b}$ is due to the transition from L as segment length to the normalized X_1.

$$\ln[\mathcal{L}_i^*(\beta_0,\beta_1,\ldots,\mathscr{b})] = \mathscr{b}X_i\ln(\mathscr{b}X_i) + [\sum_{j=1}^{m_i} k_{i,j}\ln(\widehat{E}\{\mu_{i,j}\})] + \ln\Gamma(\sum_{j=1}^{m_i} k_{i,j} + \mathscr{b}X_i) - \ln\Gamma(\mathscr{b}X_i)$$
$$-(\sum_{j=1}^{m_i} k_{i,j} + \mathscr{b}X_i)\ln[(\sum_{j=1}^{m_i}\widehat{E}\{\mu_{i,j}\}) + \mathscr{b}X_i]$$

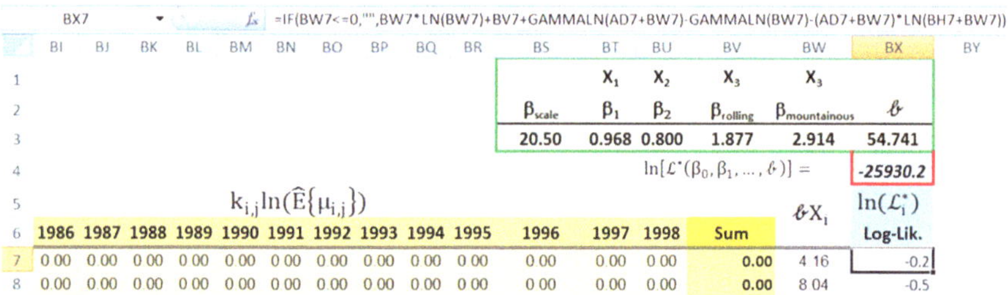
BX7 =IF(BW7<=0,"",BW7*LN(BW7)+BV7+GAMMALN(AD7+BW7)-GAMMALN(BW7)-(AD7+BW7)*LN(BH7+BW7))

	BI	BJ	BK	BL	BM	BN	BO	BP	BQ	BR	BS	BT	BU	BV	BW	BX	BY
1												X_1	X_2	X_3	X_3		
2											β_{scale}	β_1	β_2	$\beta_{rolling}$	$\beta_{mountainous}$	$\mathscr{b}$	
3											20.50	0.968	0.800	1.877	2.914	54.741	
4														$\ln[\mathcal{L}^*(\beta_0,\beta_1,\ldots,\mathscr{b})] =$		-25930.2	
5					$k_{i,j}\ln(\widehat{E}\{\mu_{i,j}\})$										$\mathscr{b}X_i$	$\ln(\mathcal{L}_i^*)$	
6	1986	1987	1988	1989	1990	1991	1992	1993	1994	1995	1996	1997	1998	Sum		Log-Lik.	
7	0.00	0.00	0.00	0.00	0.00	0.00	0.00	0.00	0.00	0.00	0.00	0.00	0.00	0.00	4.16	-0.2	
8	0.00	0.00	0.00	0.00	0.00	0.00	0.00	0.00	0.00	0.00	0.00	0.00	0.00	0.00	8.04	-0.5	

Fig. 9.12 NM parameter estimates for the 13-year panel data

9.8 Panel Data and Alternative Objective Functions

The question what to optimize, i.e., the choice of the objective function, was the subject matter of Chap. 8. Common objectives are to minimize the (weighted) sum of squared differences or to maximize likelihood. Even though the two approaches differ, when some mild conditions are met, the least-squares and maximum-likelihood parameter estimates are very similar.[46] In road safety modeling maximizing likelihood is the more popular approach. This is why it was described in detail in Chap. 8, why several likelihood functions were listed, and why, initially it was the weapon of choice.

Some alternative objective functions were examined in Sect. 8.4. One of those was the minimization of the sum of absolute differences, another that of minimizing χ^2. With the focus on fit rather than on parameter accuracy, it is simpler and more appropriate to use one of these alternative objective functions. For the C-F spreadsheet in Fig. 9.11, the minimization of the sum of |residuals| and of χ^2 yields the parameter estimates in Fig. 9.13.[47] Both result in much improved CURE plots.

---

[46] Section 8.2.1 and Charnes et al. (1976).

[47] To download this spreadsheet go to http://extras.springer.com/ and enter the ISBN of this book. Look in the "Spreadsheets" folder for Chapter 9. Alternative objectives, 13-year panel, *L*, AADT, Terrain, and common scale paramete.xls or xlsx.

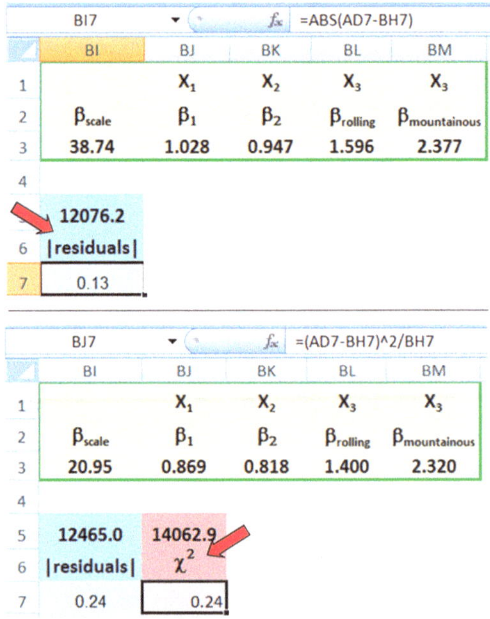

BI7 | f_x =ABS(AD7-BH7)

	BI	BJ	BK	BL	BM
1		X_1	X_2	X_3	X_3
2	β_{scale}	β_1	β_2	$\beta_{rolling}$	$\beta_{mountainous}$
3	38.74	1.028	0.947	1.596	2.377
4					
5	12076.2				
6	|residuals|				
7	0.13				

BJ7 | f_x =(AD7-BH7)^2/BH7

	BI	BJ	BK	BL	BM
1		X_1	X_2	X_3	X_3
2	β_{scale}	β_1	β_2	$\beta_{rolling}$	$\beta_{mountainous}$
3	20.95	0.869	0.818	1.400	2.320
4					
5	12465.0	14062.9			
6	|residuals|	χ^2			
7	0.24	0.24			

Fig. 9.13 Parameter estimates for two alternative objective functions

The model equation is:

$$\hat{E}\{\mu\} = \beta_{\text{scale}} X_1^{\beta_1} X_2^{\beta_2} \left(1 \text{ if } F, \beta_{\text{rolling}} \text{ if } R \text{ and } \beta_{\text{mountainous}} \text{ if } M\right) \text{acc./year}$$
$$\text{where } X_1 \equiv \text{Segment Length}/19.74, X_2 \equiv \text{AADT}/21,720 \tag{9.5}$$

For this model equation we now have three sets of parameter estimates depending on what is being minimized or maximized. That the corresponding estimates of $E\{\mu\}$ and can be markedly different is shown in Fig. 9.14.

Inasmuch as one can have very different estimates and predictions of $E\{\mu\}$ for the same data and the same model equation, the SPF is quite footloose. The explanation of why this is so and why two of the fitted curves are similar while the third is so different ties together a few points made earlier.

Recall the image of rubber bands anchored to data points and pulling on the to-be-fitted curve.[48] Because the residuals tend to be small near the origin and get larger as X_2 approaches 1, the data points on the right of the figure exert a strong pull on the to-be-fitted curve. Because the general shape of the curve is predetermined by the functional form, here by the power function $X_2^{\beta_2}$, and because the curve must go through the origin, the weak pull of the data on the left cannot resist the strong pull of the data on the right. Recall also the affinity between maximizing likelihood and

[48] See Sect. 6.6.1

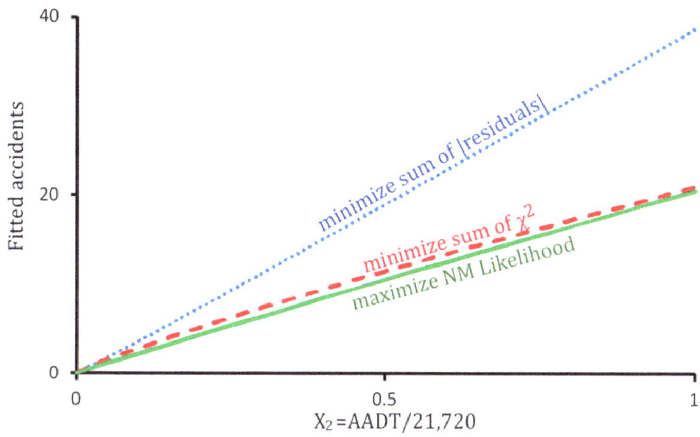

Fig. 9.14 Comparing estimates of $E\{\mu\}$ with different objective functions

minimizing the sum of squares.[49] When curve-fitting is by maximizing likelihood and minimizing χ^2 the pull of each data point is proportional to the square of its residual. However, when the sum of |residuals| is minimized the pull is proportional to the size of the residual itself. Therefore when likelihood is maximized or χ^2 minimized the position of the curve is largely dictated by the minority of data points where AADT is larger than, say, 5,000. The same is true but to a much lesser extent when the sum of absolute residuals is minimized. This is why the latter makes for a better fit for the large majority of data points on the left side of the figure.

The same reasoning explains why the CURE plot is better when the sum of absolute residuals is minimized than when likelihood is maximized. The CURE plot describes how well a model fits by cumulating residuals, not by cumulating their square. Naturally, therefore, when the objective function is made of residuals, not of their square, the CURE plot will be better.

There is an additional reason why the few data points on the right side of the figure play such a dominant role and determine the resting position of the entire fitted curve; the reason resides in the notion of functional form. Once the functional form of the model equation is chosen and thereby a certain family of curves is specified, the entire curve moves as a unit under the collective pull of the rubber bands. The changing of a parameter lowers or raises all points of curve with the origin as the fulcrum. This is a specific manifestation of the straightjacket[50] nature of model equations.

[49] See e.g. Charnes et al. (1976).

[50] See Sect. 4.2.

9.9 Adding Another Variable: Year

Terrain was said to be a proxy for traits such as grade, curvature, or roadside about which the Colorado data set is mum. Similarly, "Year" is a proxy for traits such as speed, precipitation, driving culture, road user demography, special events, or vehicle fleet which all change over time and about which data is not usually available. Now the task is to check whether in the 1986–1998 period "Year" was safety related and, if yes, to examine how Year should be added to the model equation.

A variable was said to be safety related if not including it in the model would cause bias-in-use. That this is the case for "Year" is evident from the comparison of the yearly Observed/Fitted ratios in Table 9.2. To illustrate, if Eqn. 9.5 and the parameter estimates from the top panel of Fig. 9.13 were used to obtain fitted values then, for 1986 this would be an underestimate by 19 % while for 1998 it would be an overestimate by 12 %.

The progression of these yearly ratios is shown in Fig. 9.15. The fluctuations around the trend line reflect the annual change in factors such as precipitation, economic activity, and special events. The trend line represents the ongoing changes in demography, vehicle fleet, road user culture, safer roads, better medicine, etc.

Table 9.2 Yearly Observed/Fitted ratios based on minimizing $\sum$|residuals|

	1986	1987	1988	...	1996	1997	1998
Observed	1,619	1,626	1,563	...	1,872	1,837	1,850
Fitted	1,362.1	1,362.2	1,387.6	...	1,889.8	2,007.6	2,098.9
Obs./Fit. ratio	1.19	1.19	1.13	...	0.99	0.92	0.88

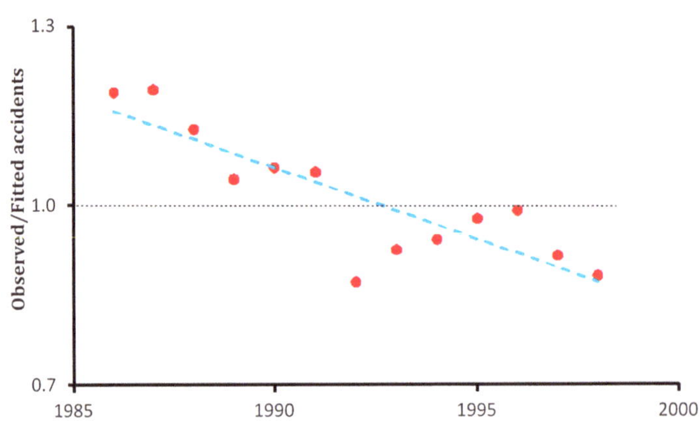

Fig. 9.15 Yearly Observed/Fitted accidents ratios

As in the case of Terrain, there are several ways in which the "Year" variable could be added to the model equation. One is to use $\hat{E}\{\mu\} = \beta_0 X_1^{\beta_1} X_2^{\beta_2} \beta_{\text{Terrain}} \beta_{\text{year}}$ with the ratios from Table 9.2 for β_{year}. To illustrate assume that the interest is in the $\hat{E}\{\mu\}$ for 1988 of a 0.5 miles long road segment of a rural two-lane road in the rolling terrain of Colorado with AADT = 500. If so $\hat{E}\{\mu\} = 38.74 \times (0.5/19.74)^{1.028} \times (500/21,720)^{0.947} \times 1.596 \times 1.13$ I&F accidents. Another way to account for the influence of "Year" would be to replace the β_{scale} by 13 parameters, one for each year. Yet another possibility is to save on parameters and multiply the model equation by a time trend factor such as $[1 - \beta_{\text{slope}}(\text{Year-}1986)]$ or to run a regression through the ratios in Table 9.2. This option would allow extrapolation.

9.10 Summary

This chapter revolved around the question of whether a variable should be added to the model equation and how to do so in a C-F spreadsheet. If when the variable is added to the model equation the estimate of $E\{\mu\}$ changes then the variable is said to be "safety related." Were a model equation without that variable used for estimating $E\{\mu\}$ when the value of that variable is known, the estimate of $E\{\mu\}$ would be biased. This is the "bias-in-use." The purpose of adding a variable to the model equation is to reduce the bias-in-use. It follows that the necessary conditions for the addition of a variable are two: (1) that the variable be "safety related," i.e., that introducing it into the model equation will reduce bias-in-use and (2) that information about the variable be usually available for applications.

These conditions are "necessary" but are not "sufficient." When a variable is added to the model equation there is a gain and a loss. The gain is that of reducing the bias-in-use. The loss is in that the addition of a variable often increases the variance of the estimate of $E\{\mu\}$. The add-or-do-not-add decision involves a balancing of these two conflicting tendencies. Doing so is not simple.

Whether a candidate variable is safety related has to be examined just prior to its addition to the model equation. This requires the conduct of a Variable Introduction Exploratory Data Analysis – VIEDA. Thus, for example, to determine whether AADT should be added to the model equation after the influence of Segment Length was accounted for, the relationship between the ratio of Observed Accidents/Fitted Accidents and AADT was examined. As expected, the ratio indicated the existence of a regular pattern. In this manner AADT was confirmed to be safety related and amenable to introduction into the model equation by a smooth function. A similar VIEDA preceded the addition to the model equation of the Terrain variable and of Year.

The addition of a variable to the model equation requires a modification of the C-F spreadsheet following which all parameters are reestimated. As a result previously estimated parameter values usually change. It follows that whatever the parameter estimates are at some stage of modeling, they must be viewed as provisional and biased. The bias is that caused by the variables not in the model equation at that stage – the Omitted Variables Bias. If a new safety-related variable was added to the model equation, some parameter estimates would change again. For this reason, little trust should be placed in the statistics that are commonly used to describe the accuracy of parameter estimates.

To this point the average AADT and the sum of accidents for a single 5-year period served as data. However, the original Colorado data contain information for accident counts and AADTs in each of 13 years. The NB likelihood function pertains to single-period data. In contrast, the NM likelihood function allows the use of yearly "panel" data and has been implemented in this chapter. Its chief advantages are two. First that it makes use of all the information available. Second, that it allows the introduction of time as a variable.

The process of model development is that of a gradual build-up. In this chapter AADT, Terrain, and Year variables were added to Segment Length one after another. Rather than reporting only on the final fully loaded model, the estimated SPF was reported at each stage. This practice allows the user to choose the SPF for which he or she has the data.

The model equation is an algebraic expression made up of variables, parameters and algebraic operations.[51] When a variable is to be added to the model equation the modeler has to decide how to weave it into the fabric of the existing algebraic expression. So far the assumption was that each new variable will be represented by a function that multiplies the existing expression and that the new function would be either a constant, as in the case of Terrain and Year, or a power function X^{β} for the Segment Length and AADT variables. These provisional assumptions were made for clarity in exposition, because there is no theory to guide the choice of a model equation, inasmuch as these assumptions are common practice, and because the EDA does not support finer distinctions or more elaborate assumptions. However, there is no reason to think that these simplistic assumptions and choices make for a good fit and accurate estimates or predictions. The difficult question of how to choose the function behind the data is the theme of the next chapter.

[51] Algebraic operations are: addition, subtraction, multiplication, division and exponentiation.

References

Allen MP (1997) Understanding regression analysis. Springer, New York
Charnes A, Frome EWL, Yu PL (1976) The equivalence of generalized least squares and maximum likelihood estimates in the exponential family. J Am Stat Assoc 71(353):169–171
Draper N, Smith H (1981) Applied regression analysis, 2nd edn. Wiley, New York
Gross F, Craig L, Persaud B, Srinivasan R (2013) Safety effectiveness of converting signalized intersections to roundabouts. Accid Anal Prev 50:234–241
Harwood DW, Council FM, Hauer E, Hughes WE, Vogt A (2000) Prediction of the expected safety performance of rural two-lane highways. FHWA-RD-99-207. Office of Safety Research and Development, Federal Highway Administration, Washington, DC
Hauer E (2004) Statistical safety modelling. Transportation Research Record 1897. National Academies Press, Washington, DC, pp 81–87
Lehmann ER (1990) Model specification: the views of Fisher and Neyman, and later developments. Stat Sci 5(2):160–168
Lord D (2008) Methodology for estimating the variance and confidence intervals for the estimate of the product of baseline models and AMFs. Accid Anal Prev 40:1013–1017
Lord D, Kuo P-F, Geedipally SR (2010) Comparison of application of product of baseline models and accident-modification factors and models with covariates. Transportation Research Record. No. 2147, pp 113–122
Wood GR (2005) Confidence and prediction intervals for generalized linear accident models. Accid Anal Prev 37:267–273
Zegeer CVD, Reinfurt DW, Hummer J, Herf L, Hunter W (1988) Safety effects of cross-section design. Transportation Research Record 1195. National Academies Press, Washington, DC, pp 20–32

Choosing the Function Behind the Data 10

Abstract
The general form of the model equation is $E\{\mu\} = f(X_1, X_2, \ldots, \beta_0, \beta_1, \ldots)$ where $f()$ stands for some algebraic expression. For modeling one has to choose some specific $f()$ and the question was how to do so. Even if the right $f()$ is not known one can still make decent estimates of $E\{\mu\}$. However, to say what change in $E\{\mu\}$ is caused by a some change in a predictor variable, not knowing the right $f()$ can be a problem. The task of fashioning the $f()$ out of predictor variables and parameters is difficult. Road safety data can be represented by many different and footloose functions, and what the modeler chooses matters in terms of what estimates are produced. Limited guidance is offered.

10.1 The Holy Grail

The generic model equation is $E\{\mu\} = f(X_1, X_2, \ldots, \beta_0, \beta_1, \ldots)$ where $E\{\mu\}$ is the "dependent variable" and $f()$ is some algebraic function of the "predictor variables" $X_1, X_2, \ldots$ and of the parameters $\beta_0, \beta_1, \ldots$. Determining what $f()$ hides behind noisy data is the holy grail of modeling. If we knew $f()$ then a good fit would be the promise of good estimates and good predictions. More importantly, it would remove one of the main obstacles to the cause-effect interpretation of SPFs. It is therefore peculiar that in both research and in practice so little attention is given to what seems to matters most – the function in which the predictor variables combine. Without explanation or comment most modelers tend to assert that some simple linear or log-linear expression *is* the model equation and then proceed to estimate the βs as if what matters are the parameters, not the function to which they belong. Vexing is also the paucity of tools available to modelers for choosing a good model equation.

Parameters are only the plumage on the body of the model equation; they have no substance without the model equation and are of little interest if they do not belong to the correct one. To illustrate, one could assume, along with Aristotle and

E. Hauer, *The Art of Regression Modeling in Road Safety*,
DOI 10.1007/978-3-319-12529-9_10

all his followers, that the acceleration of a free falling body is proportional to its weight. With such a model equation in mind one could use data to estimate the missing parameter – the constant of proportionality. As Galileo demonstrated, such a parameter belongs to the wrong model and therefore is of no use nor meaning. The choice of the model equation is the primary activity, and the subsequently estimated parameters depend for their meaning and usefulness on it.

It would be good if there were logical grounds or theory-based hypotheses to guide the choice of the model equation in road safety; if it could be chosen for reasons more weighty than habit and simplicity. A hesitant first step towards a theory is in Appendix J. Alas, at this time, convention, goodness of fit, parsimony of parameters, and, occasionally, performance in estimation, seem to be the only guides for choosing the model equation.

There are very many model equations to choose from and little to guide the choice.

As illustrated in the next section, even when *f*() is simple and the data good, the function can be difficult to identify. This puts the road safety modeler in a tough position. On one hand there is the absence of theory and thus the necessity to somehow extract a good model from data. On the other hand there is the harsh reality of an unknown and possibly complex relationships hiding within the cloud of noisy and incomplete cross-sectional data. Like Tantalus, modelers keep reaching for a fruit that is forever receding.

10.2 The Elusive *f*(): A Story with Morals

Imagine a researcher who, walking on the Alexandria's seashore, found some regular looking flat objects. The researcher wanted to determine how the longest dimension of these objects, Y, depends on their shorter dimensions, X_1 and X_2. The researcher measured the Y, X_1, and X_2 of 100 such objects. The measurement error was normally distributed, $\mathcal{N}(\mu = 0 \text{ m}, \sigma = 0.01 \text{ m})$. When in the course of the EDA the researcher plotted the measured Ys against the X_1s and the X_2s Fig. 10.1 obtained.

Looking at the figure the researcher thought that the linear-additive form $Y = \beta_0 + \beta_1 X_1 + \beta_2 X_2$ might be satisfactory. Using the measured data[1] the estimates of the βs were 0.35, 0.73, and 0.66.[2] The researcher did not know that the

[1] Hundred data points were generated such that the X_1 are uniformly distributed in between 2 and 10 m, the X_2 between 3 and 8 m, the hypotenuse Y was computed by the Pythagorean theorem, and a $\mathcal{N}$ (0, 0.01 m) random error was added to each measurement.

[2] Seeing the objects and having measured their dimensions the researcher could have reasoned that if both X_1 and X_2 are very short Y must also be so and therefore, the intercept β_0 should be set to 0. Doing so would remove a logical blemish but make for a slightly worse fit.

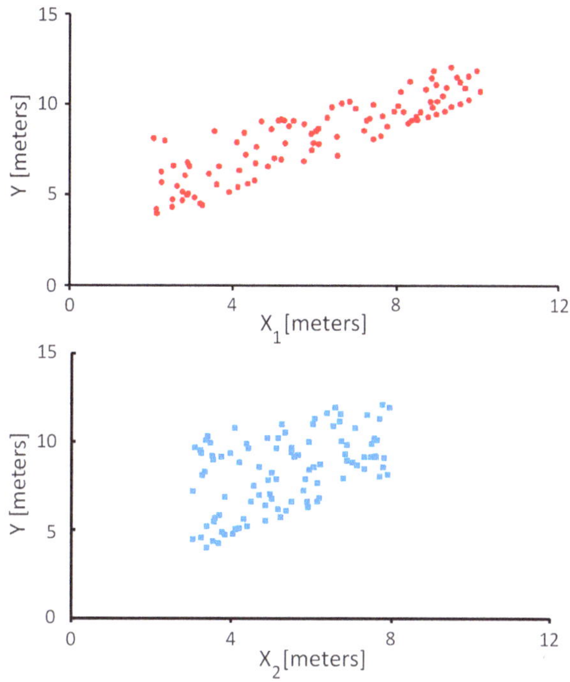

Fig. 10.1 What the researcher sees

Pythagorean theorem applies to these objects and that the model $Y = \left(X_1^{\beta_1} + X_2^{\beta_2}\right)^{\beta_3}$ would have been the right choice of *f*().[3]

The moral of this story is manifold. First, as noted earlier, parameters belong to what is chosen to be the model equation. The *β*s of the linear-additive model have nothing to do with the *β*s of the Pythagorean model. Since the meaning and magnitude of the *β*s depends on the *f*(), and not the other way around, the task of choosing and shaping the function *f*() is of primary importance in modeling; parameter estimation is secondary.

The second lesson is that the true functional form of the model equation is nearly unfathomable. Is very unlikely that the use of the Pythagorean functional form would have occurred to many modelers had they not already known of its applicability to the data, in which case no statistical modeling would have been necessary;

[3] Were the Pythagorean model used the estimates of β_1, β_2, and β_3 for the same data would have been 1.98, 1.98 and 0.50. That is, if one had the right *f*(.) then regression would yield the right parameters.

logic and reasoning would do. I asked dozens graduate students to fit a model to this kind of data; not one discovered the Pythagorean theorem by curve-fitting.

The Pythagorean theorem cannot be discovered by regressions.

Third, the linear-additive form chosen by the modeler is in some sense "simpler"[4] than the Pythagorean form. It follows that the guidance of Occam's razor[5] that counsels to prefer simpler models is not always good advice. It this example, and perhaps in most representations of the real world by a model, systematic preference for simplicity may be to the detriment of good modeling.

Fourth, however reasonable the choice of the linear-additive form in this case, its parameter estimates tell little about how Y would change if X_1 or X_2 were manipulated. It would be not only inaccurate but entirely indefensible to think that if X_1 was prolonged by 1 m then Y would increase by β_1(=0.73 m), as the regression equation suggests. When X_1 is small compared to X_2 the increase in Y would be close to 0 m, whereas when X_1 is large compared to X_2 the increase in Y would be close to 1 m. In reality, the amount of change in Y depends on the sizes of both X_1 and X_2 while the linear form $Y=\beta_0+\beta_1X_1+\beta_2X_2$ erroneously implies that the change in Y due to a change in an X is a constant.[6]

Fifth, as is shown in Fig. 10.2, the correspondence between the observed values of Y and those estimated by the linear-additive model is very good. Still, the linear-additive model fails to predict the effect on Y of manipulating an X. It follows that the hope that a well-fitting model has a good chance of predicting the effect of manipulations is just that – a hope. One can use the regression equation to say fairly accurately how long is Y when X_1 and X_2 are known but one cannot reliably use it to say by how much Y would be changed if X_1 or X_2 were made longer or shorter. Translated into the road safety context, when the estimate of the $E\{\mu\}$ is to serve the practical uses discussed in Sects. 1.3 and 1.4 one does not have to have the right *f*(). However, to say how $E\{\mu\}$ would be altered if some predictor variable was manipulated, having the right *f*() is of essence.[7] Without the right *f*() the cause-and-effect message of the regression equation cannot be trusted.

---

[4] The concept of simplicity is an esthetic one and has no agreed upon measure.

[5] William of Ockham's (c. 1287–1347) razor says something like: "All things being equal, the simplest solution tends to be the best one." The "tends to be" means that simplest is not always the best. Besides, what makes the linear model "simpler"? Some object to Ockham's razor, because it seems to imply that of two alternative representations of nature the simpler one is more likely to be true. There is no empirical evidence for such a belief.

[6] This leads to another important point which will be elaborated on in Sect. 10.9, namely, that by choosing the functional form the modeler predetermines how the dependent variable will change due to a change in a predictor variable.

[7] The same distinction was made in Sect. 6.7 where the purposes of regressions were discussed. The three purposes were: (1) to summarize or to describe a body of data; (2) to predict the value of a dependent variable Y (here $E\{\mu\}$) from a set of independent variables $X_1, \ldots, X_n$; (3) To predict the change in the value of the dependent variable from an intervention that changes the value of X_1, etc. The legitimacy of purpose (3) is widely questioned.

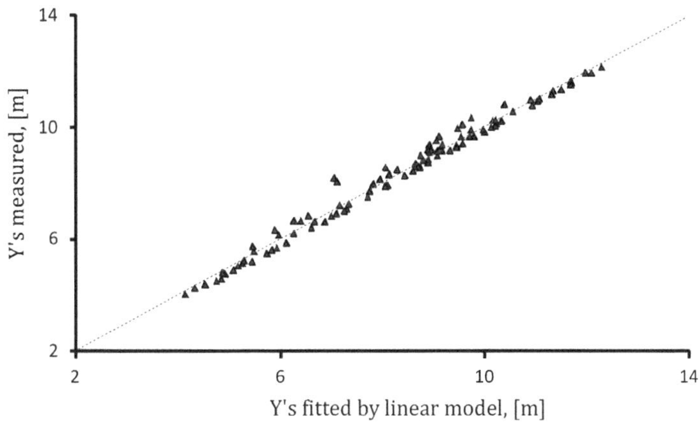

Fig. 10.2 Correspondence between observed and estimated *Y*s for the linear-additive model

The task of identifying the right *f*() from data is a tall order.

This brings us back to the two perspectives discussed in Sect. 1.5. When the focus is on practical applications (e.g., screening for blackspots, determining cost effectiveness of treatments, etc.) what matters is the quality of the estimates of $E\{\mu\}$ and $\sigma\{\mu\}$. In this case there is no reason to question odd looking model equations and parameter estimates. Only when the regression equation is to be used to compute the effect on $E\{\mu\}$ of a change in a predictor variable is an odd looking model and parameter estimate disconcerting.[8]

In this story the researcher chose to estimate the parameters of the linear-additive model form. This choice was reasonable because looking at Fig. 10.1 the impression is that Y increases by equal amounts for equal increases of X_1 and the same can be said, albeit with less conviction, about the association of Y and X_2. This is consistent with the linear-additive form which leads to the incorrect conclusion that adding 1 m to X_1 increases Y by β_1 and adding 1 m to X_2 increases Y by β_2. The researcher could have easily chosen the popular multiplicative form $Y = \beta_0 X_1^{\beta_1} X_2^{\beta_2}$ which has a more nuanced relationship between ΔX and ΔY.[9] Such a model equation might also be consistent with the data clouds in Fig. 10.1 although its

[8] I always found Henri Theil's comforting aphorism that "Models are to be used, not believed" puzzling; can one use a model without believing that what it tells us is approximately true? Perhaps what Theil (1971) meant was that one need not believe that the function and its parameters are right if the predictions produced by the model are close to the mark. In the road safety context Theil's cryptic statement applies only to practical applications perspective. If the model is to be used to tell by how much $E\{\mu\}$ is likely to change as a results of some change in a predictor variable, one must believe to have the substantively correct function and parameters.

[9] The differential is $dY = \frac{Y}{X_i}\beta_i dX_i$.

presence in the data is not visually indicated. Had the researcher opted for the multiplicative form without trying the linear-additive form first, the correspondence between the observed and fitted values would still be quite impressive and the multiplicative model also would have been found satisfactory. A nearly perfect fit could be achieved had the researcher tried the seemingly arbitrary model $Y = \beta_0 X_{\text{Max}}\left[1 + \beta_1\left(\frac{X_{\text{Max}}}{X_{\text{Min}}}\right)^{\beta_3}\right]$ in which X_{Max} is the larger of X_1 and X_2 and X_{Min} the smaller one. Nobody would come up with this model equation on the basis of an EDA and its data visualizations.

EDA, visualization, and habit incline the mind towards the use of linear and multiplicative (log-linear) model forms. However, as this story makes clear, when the underlying reality is neither linear nor multiplicative, as in the case of the hypotenuse of right angle triangles, the model equation cannot be used to predict the effect of manipulations. For this purpose even a good correspondence between observed values and estimates is of no comfort.

10.3 Enroute to the Multiplicative Model Equation

The subject of road safety modeling are the complex interactions between the road, the traffic flow on it, the characteristics of road users and their vehicles, maintenance procedures and budgets, the law and its enforcement, the climate, etc. This is the complexity the substance of which the model equation should aim to capture and represent. However, as has been illustrated, even simple functions of few variables may be difficult to identify.

While the general model equation is $E\{\mu\} = f(X_1, X_2, \ldots, \beta_0, \beta_1, \ldots)$ with $f()$ unknown, most modelers are content to declare that a function such as $E\{\mu\} = e^{\sum_{\forall i}\beta_i X_i}$ *is* the model equation.[10] What justifies the abandonment of the quest for the unknown function $f()$ and its replacement by a simple learned-looking assertion? The best discussion of the topic that I know about is by Lau (1986) who explains the transition from the unknown $f()$ to the postulation of a specific function. The transition consists of a series of steps.

The first step in the sequence is to narrow the domain of all possible functions $f()$ to only those that are linear in parameters.[11] This, Lau explains, was done for historical reasons and for convenience in parameter estimation. With this, $f(X_1, X_2, \ldots, \beta_0, \beta_1, \ldots)$ is replaced by $\sum f_i(X_1, X_2, \ldots)\beta_i$. To restrict the domain further Lau says that "it is often desirable to be able to identify the effect of each

[10] See, e.g., Eq. (2) in Shankar et al. (1995), Eq. (2) in Poch and Mannering (1996), Eq. (3) in Lord and Bonneson (2007), Eq. (4) in Chiou and Fu (2013), etc.

[11] The model $Y = \beta_0 + \beta_1 X_1 + \beta_2 X_2 + \cdots$ is linear in both parameters and variables. The model $Y = \beta_0 + \beta_1\sqrt{X_1 + X_2}$ is linear in parameters but not in variables. The model $Y = \beta_0\sqrt{X_1^{\beta_1} + X_2^{\beta_2}}$ is not linear in either parameters or variables.

independent variable on the dependent variable separately." (p. 1517).[12] With this $\sum f_i(X_1, X_2, \ldots)\beta_i$ is replaced by $\sum f_i(X_i)\beta_i$. The next step in the chain of domain restrictions and function simplifications is to assume that, the function $f_i(X_i)$ which heretofore may have differed from one predictor variable to another is, in fact, the same function $f(X_i)$ for all predictor variables. This is done "for ease of computation and interpretation and for aesthetic reasons." (p. 1517) In this way, after a series of domain restrictions motivated history, estimation convenience, ease of computation, and aesthetics, one ends up with $E\{\mu\} = \sum f(X_i)\beta_i$ or some order-preserving transformation thereof. The last step is to choose something specific for $f(X_i)$. If, for example, one declares that $f(X_i) = X_i$ then the linear-additive model equation obtains. If, in addition, one opts for the order-preserving transformation $\ln[E\{\mu\}] = \sum X_i\beta_i$ then the multiplicative model equation $E\{\mu\} = e^{\sum_{\forall i}\beta_i X_i}$ mentioned earlier obtains.

Lau notes that such simplified functional forms "may be interpreted as first-order approximations to any arbitrary function" in the vicinity of a specific point and "that is one reason why they have such wide currency." (p. 1517). That is, if one knows the value of a function and its slope at some point, one can compute its approximate value at a nearby point. This works well if the slope between the two points is nearly constant. In road safety one may not assume that the slope of the SPF remains nearly constant for the usually encountered ranges of traffic flow, sight distance, sideslope, etc. This is why, in road safety modeling, the approximation argument does not justify the use of models based on $\sum X_i\beta_i$. Apparently the same is often true in econometrics about which Lau says that "... linear functions, while they may approximate whatever underlying function reasonably well for small changes in the independent variables, frequently do not work very well for many others purposes" (pp. 1517–1518).

Model equations in road safety tend to follow all the aforementioned restrictions of function domain and of simplification. Commonly used are multiplicative models made of single-variable building blocks such as $E\{\mu\} = (\text{Segment Length}) \times \left(\beta_0 X_i^{\beta_1} X_2^{\beta_2} \ldots\right)$ or $E\{\mu\} = (\text{Segment Length}) \times \left(e^{\beta_0} e^{\beta_1 X_1} e^{\beta_2 X_2} \ldots\right)$. However, not all the simplifications which, as Lau explains, have historical roots or convenience as justification are still necessary. Even if the multiplicative structure is retained, not all function in the product need to be the same. Neither do the functions have to be linear in parameters nor precast as $\beta_1 X_1, \beta_2 X_2, \ldots$ or $X_1^{\beta_1}$, $X_2^{\beta_2}, \ldots$ or $e^{X_1\beta_1}, e^{X_2\beta_2}, \ldots$; the functions can take on any form. Accordingly, in what follows the generic form of (10.1) will be used.

[12] Why exactly is it desirable to make f_i a function of a single predictor variable? According to Lau doing so allows the parameter β_i to be interpreted as a measure of the effect on the dependent variable of a change in the predictor variable X_i. This convenience-motivated simplification is an assumption about reality. In road safety it means, e.g., the safety effect of lane widening does not depend on traffic flow, shoulder design, or any other predictor variable.

$$E\{\mu\} = \beta_0 \times f_1(X_1, \boldsymbol{\beta}_1) \times f_2(X_2, \boldsymbol{\beta}_2) \times \cdots \times (\text{interaction terms}) \qquad (10.1)$$

Here β_0 is a scale parameter; f_1 is a function of X_1, f_2 is a function of X_2, etc.; $\boldsymbol{\beta}_1$ is the set of parameters $\beta_1, \beta_2, \ldots, \beta_n$ in f_1, $\boldsymbol{\beta}_2$ is the set of parameters $\beta_{n+1}, \beta_{n+2}, \ldots$ in f_2, etc. In addition, when appropriate, interaction terms consisting of several variables and parameters will be added.[13] This multiplicative model form is still much more limiting than the general function $E\{\mu\} = f(X_1, X_2, \ldots, \beta_0, \beta_1, \ldots)$.[14] However, (10.1) liberates modeling from near dogmatic adherence to $E\{\mu\} = e^{\sum_{\forall i} \beta_i X_i}$ and the difficult-to-justify, and largely unnecessary restrictions which it embodies.

10.4 Trying for a Better Fit

At this point the model equation is $\hat{E}\{\mu\} = \beta_0 X_1^{\beta_1} X_2^{\beta_2} \beta_{\text{Terrain}} \beta_{\text{year}}$. By the notation of (10.1), $f_1(X_1, \boldsymbol{\beta}_1) = X_1^{\beta_1}$, $f_2(X_2, \boldsymbol{\beta}_2) = X_2^{\beta_2}$, $f_3(\text{Terrain}, \boldsymbol{\beta}_3) = \beta_{\text{Terrain}}$, etc. This functional form – a product of power functions and its parameters – was to be a starting point; it was chosen for its simplicity and because the EDA did not allow for finer distinctions. However, it would be surprising if the complex process that generated the Colorado data could be well approximated by such a simple expression. The baseline CURE plots for this model[15] are in Fig. 10.3. Indeed, as is obvious, they are not acceptable. The question is how to improve an inadequate model equation.

Two problems are manifest in Fig. 10.3. There is a precipitous drop between points A and B and there is a long decline between points C and D. Different problems may require different remedies.

10.4.1 Remedy I: A Bump Function for Segment Length

A vertical drop in the CURE plot may conceal an outlier. However, as shown in Fig. 10.4 the drop between A and B is really an agglomeration of many small residuals; there are no outliers there. By Table 10.1, in the Colorado data there are 373 segments that are 0.89 and 0.90 miles long and on these the fitted accidents consistently exceed the observed accidents.

To eliminate the A to B drop one could multiply the $\hat{E}\{\mu\}$ in (9.5) by 0.79 when $X_1 = 0.89/19.74$ and by 0.73 when $X_1 = 0.90/19.74$ miles. However, the deliberate

[13] The subject of interaction will be discussed in Sect. 10.9.

[14] One of the main limitation of the multiplicative model is that road segments are never homogeneous in their traits. The presence of a narrow bridge, a horizontal curve or of a driveway affects the safety of only a part of the segment and as such should be accounted for by an addition, not a multiplication. For discussion see Hauer (2004).

[15] Specifically, the fit minimizing $\sum|\text{residuals}|$ from Fig. 9.13 was used.

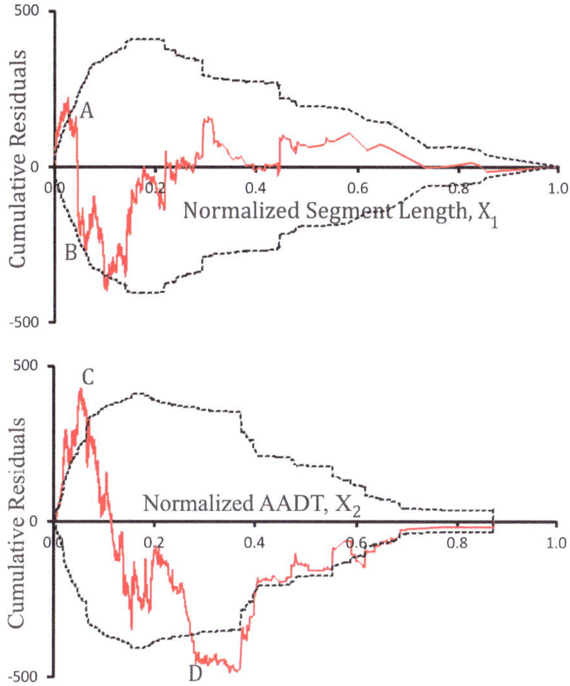

Fig. 10.3 Baseline CURE plots

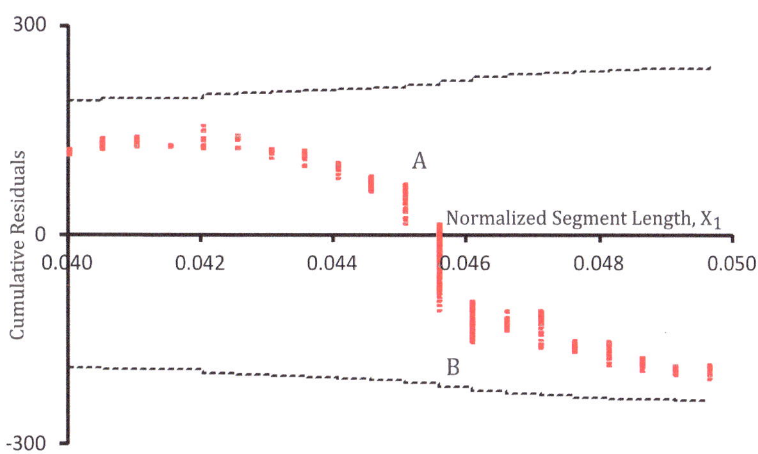

Fig. 10.4 CURE plot details

Table 10.1 Data near the A to B drop

Segment length (miles)	X_1	No. of segments	Observed accidents	Fitted accidents	Observed/ fitted
0.89	0.0451	117	203	257.85	0.79
0.90	0.0456	256	312	429.38	0.73

introduction of a discontinuity in the continuous function $X_1^{\beta_1}$ may seem objectionable as we are inclined to think that the response to a continuous variable should be free of jumps. It would be particularly displeasing to those who see in regressions a tool for the discovery of causal links. However, the existence of the drop in the CURE plot merely indicates that segments that are 0.89 and 0.90 miles long are different in some way from other segments, and that this difference is not accounted for in the model equation. Were it known what traits sets segments of this length apart from other segments, and were these traits appropriately represented in the model equation, there should be no drop in the CURE plot and it would not be necessary to use multipliers.

Seen in this light the multipliers do not signify an objectionable discontinuity in an otherwise continuous function and underscores the fact that the assumed continuous function $X_1^{\beta_1}$ is not some kind of natural law; it is only a compact rule for computing estimates of $E\{\mu\}$. That something can be succinctly described by an algebraic expression does not endow it with the prestige and mystique that goes with mathematical logic or the rigor of science. Nor does it promise "transferability" – the applicability of the fitted relationship beyond the data for which it was obtained.

Instead of applying two multipliers and enshrining a discontinuity in the model equation, one could apply a continuous patch to a wider region.[16] Let X_L denote the lower limit of the range of variable X to which the Bump Function patch is to apply and X_U its upper limit. Define

$$t \equiv \frac{X - X_L}{X_U - X_L} \quad \text{when} \quad X_L < X < X_U \tag{10.2}$$

A convenient bump function (BF) is

$$\begin{aligned} \text{BF} = &\left[1 + t^{\beta_{BF,1}}(1-t)^{\beta_{BF,2}}\right] \text{ when the fitted value is too small and} \\ &\left[1 - t^{\beta_{BF,1}}(1-t)^{\beta_{BF,2}}\right] \text{ when the fitted value is too large,} \\ &\text{where } \beta_{BF,1} > 0 \text{ and } \beta_{BF,2} > 0 \end{aligned} \tag{10.3}$$

How to choose a continuous bump function is shown in Appendix K. Both the two discrete multipliers and the continuous patch can eliminate the A to B drop. However, the resulting CURE plot for Segment Length remains unsatisfactory.

[16] For detail see Appendix K.

10.4.2 Remedy II: Alternative Functions

The problem with the baseline CURE plot for AADT is that between C and D the fitted values are too large and to the right of D they are too small. To determine what correction could be applied the VIEDA methods of Sect. 9.2 can be used to advantage. How the observed/fitted accidents ratio depends on the normalized AADT is shown in Fig. 10.5.

The progression of points in Fig. 10.5 suggests that multiplying the power function $X_2^{\beta_2}$ by the straight line $(1+\beta_3 X_2)$ with $\beta_3 > 0$ might help. This requires a modification of the corresponding factor in the C-F spreadsheet as is shown in Fig. 10.6.[17]

This modification improved the CURE plot substantially. The question is whether some other function of X_2 (≡AADT/21,720), one that is not just a modification of the power function, would be a better choice. To choose intelligently the modeler needs to know what various functions look like and how their shape is affected by changes in their parameters.

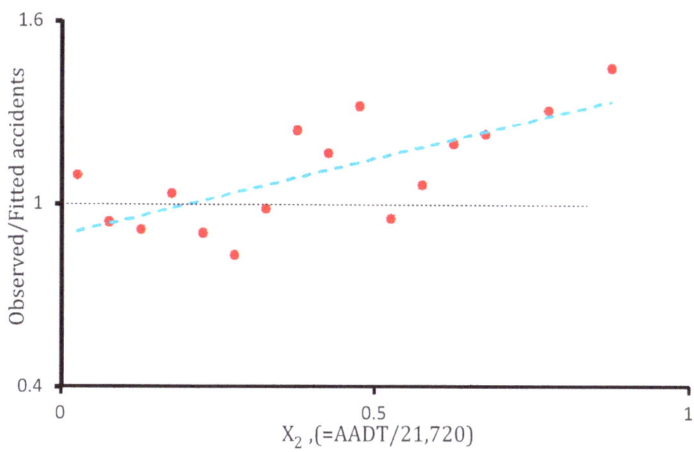

Fig. 10.5 Seeking a modification for the $X_2^{\beta_2}$ function

[17] To download these spreadsheets go to http://extras.springer.com/ and enter the ISBN of this book. Look in the "Spreadsheets" folder for "Chapter 9, Alternative objectives, 13-year panel data, L, AADT, Terrain, Common scale parameter and Chapter 10 Absolute residuals, Power x Line.xls or xlsx."

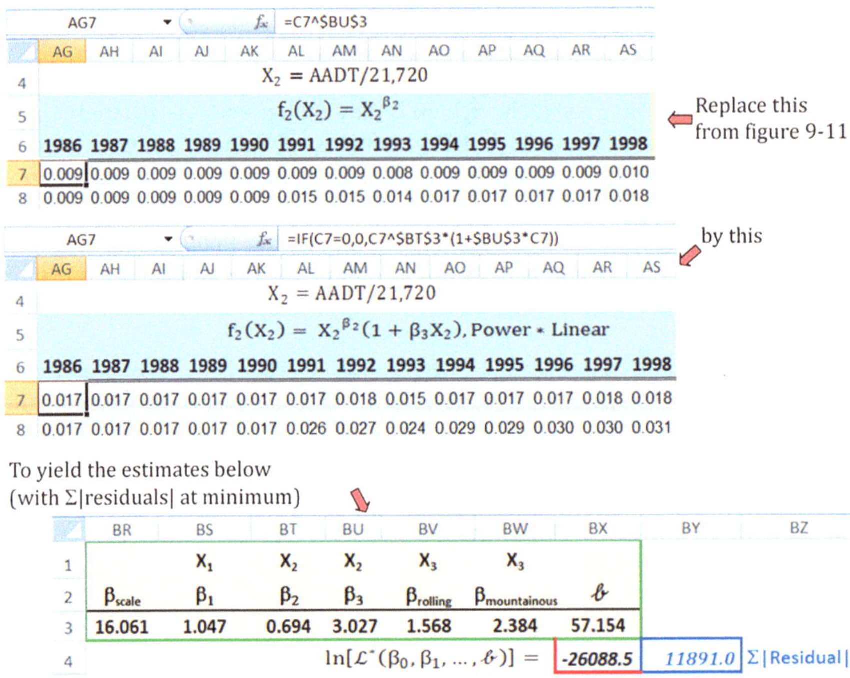

AG7 =C7^$BU$3

	AG	AH	AI	AJ	AK	AL	AM	AN	AO	AP	AQ	AR	AS
4	X_2 = AADT/21,720												
5	$f_2(X_2) = X_2^{\beta_2}$												
6	1986	1987	1988	1989	1990	1991	1992	1993	1994	1995	1996	1997	1998
7	0.009	0.009	0.009	0.009	0.009	0.009	0.009	0.008	0.009	0.009	0.009	0.009	0.010
8	0.009	0.009	0.009	0.009	0.009	0.015	0.015	0.014	0.017	0.017	0.017	0.017	0.018

AG7 =IF(C7=0,0,C7^$BT$3*(1+$BU$3*C7))

	AG	AH	AI	AJ	AK	AL	AM	AN	AO	AP	AQ	AR	AS
4	X_2 = AADT/21,720												
5	$f_2(X_2) = X_2^{\beta_2}(1+\beta_3 X_2)$, Power * Linear												
6	1986	1987	1988	1989	1990	1991	1992	1993	1994	1995	1996	1997	1998
7	0.017	0.017	0.017	0.017	0.017	0.017	0.018	0.015	0.017	0.017	0.017	0.018	0.018
8	0.017	0.017	0.017	0.017	0.017	0.026	0.027	0.024	0.029	0.029	0.030	0.030	0.031

	BR	BS	BT	BU	BV	BW	BX	BY	BZ
1		X_1	X_2	X_2	X_3	X_3			
2	β_{scale}	β_1	β_2	β_3	$\beta_{rolling}$	$\beta_{mountainous}$	b		
3	16.061	1.047	0.694	3.027	1.568	2.384	57.154		
4				$\ln[\mathcal{L}^*(\beta_0, \beta_1, \ldots, b)] =$			-26088.5	11891.0	Σ|Residual|

Fig. 10.6 Modifying the "Power" function

10.5 What Equations Look Like

Some can look at a mathematical expression and see the shape of the corresponding graph in their mind's eye; they can also envision how the changing of a parameter alters the function's shape. Most do not have this skill. For those a "visualization spreadsheet tool" was created.[18] Programmed into this tool are the four "basic" and three "modifier" functions listed in Table 10.2.

The "basic" functions may represent the dependence of $E\{\mu\}$ on "basic variables" such as Segment Length and AADT that start at the origin and take on positive values.[19] The "modifier" functions, in contrast, start with the ordinate of 1 at $X = 0$. Multiplying a basic function by the modifier serves two purposes. One is to add detail to the shape of the "basic" function, the other is to represent the influence of variables such as "Degree of Curve" or "Lane Width" which modify (increase or decrease) the $E\{\mu\}$.

[18] To download this spreadsheet go to http://extras.springer.com/ and enter the ISBN of this book. Look in the "Spreadsheets" folder for "Chapter 10. The visualization tool.xls or xlsx."

[19] The addition of a positive intercept can raise the entire basic function by a fixed amount.

For the parameter values in Fig. 10.7 the Visualization Tool Spreadsheet shows the shapes of the basic functions in Fig. 10.8 and of the modifier functions in Fig. 10.9. By changing these parameters on the spreadsheet one can observe the response in the graphs.

Table 10.2 Factors for multiplicative model equations

Basic	Modifier
1. Power: X^{β_1}	5. Exponential: $e^{\beta_1 X}$
2. Polynomial: $\beta_1 X + \beta_2 X^2 + \beta_3 X^3 \ldots$	6. Linear: $1 + \beta_1 X$
3. Logistic: $1/\left(1 + \beta_1 e^{\beta_2 X}\right) - 1/(1 + \beta_1)$	7. Quadratic: $1 + \beta_1 X + \beta_2 X$
4. Weibull: $1 - e^{-(X/\beta_1)^{\beta_2}}$	

	A	B	C	D	E	F	G	H
25			Basic				Modifier	
26		1.Power	2. Polynomial	3. Logistic	4. Weibull	a. Exponential	b. Linear	c. Quadratic
27	β1=	2	0.2	9	1	-1.5	-0.3	0.1
28	β2=		-0.2	-2	2			-0.06
29	β3=		0.1					

Fig. 10.7 Parameters used in Figs. 10.8 and 10.9

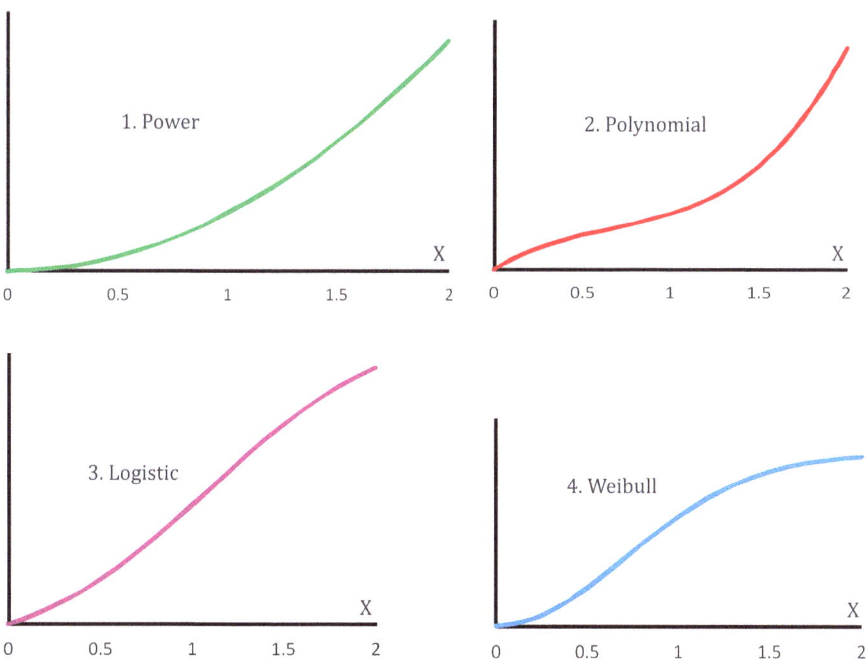

Fig. 10.8 Graphs of some basic functions

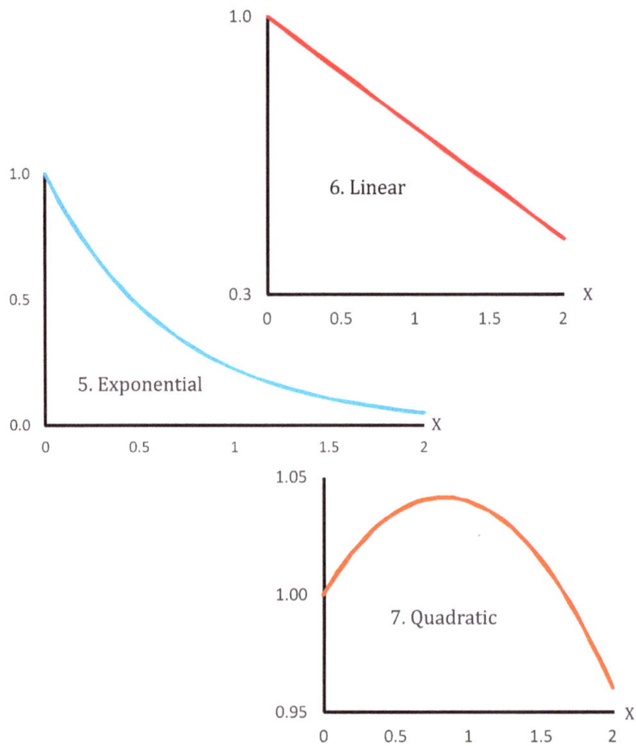

Fig. 10.9 Graphs of some modifier functions

Combining three basic and four modifier functions creates 12 potentially useful "composite" functions. Thus, for example, multiplying the basic = power function by the modifier = exponential function yields the Hoerl (1950) composite function in Fig. 10.10.[20]

Equipped with this tool and understanding one can make modifications to the model equation in (9.5) attempting to improve the fit.

[20] The Hoerl function is used mainly for curve-fitting. It arises in physics where it represents the dependence of the strong nuclear force on the distance between particles and, in inverse form, in the vapor pressure curve. However, here it is not representing any theory. The most that can be said is that if the embryonic theory in Appendix J has some substance then the Hoerl function could approximate the general shape which that theory suggests.

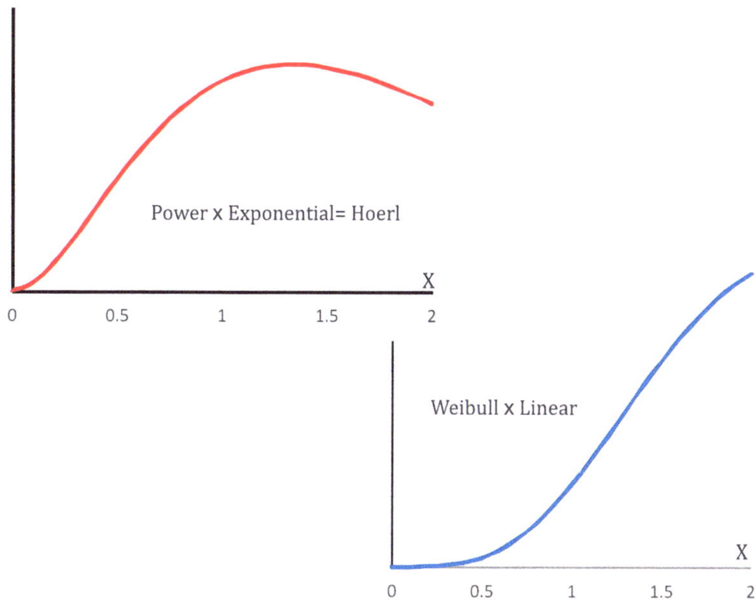

Fig. 10.10 Combining basic and modifier functions

10.6 Trying Various Functions

The functions of X_2 tried, along with their parameter estimates are listed in Table 10.3 and the corresponding graphs are in Fig. 10.11.

Note that up to about $X_2 = 0.4$ (AADT $\cong$ 9,000) the six functions in groups A and B are practically indistinguishable. Here it makes no difference which function from this repertoire is chosen. For larger values of X_2 the curves diverge such that within each group of three the estimates of $E\{\mu\}$ are nearly overlapping whereas between the two groups the differences are practically important.

It is remarkable that functions which are so different in appearance can very nearly overlap in the range containing more than 90 % of the data segments. This is also the range where the CURE plot of Fig. 10.3 is deficient; so much for the hope that one of these alternative functions will remedy this bias-in-fit problem.

Noteworthy is the fact that the six functions in groups A and B all have very similar log-likelihoods. How then to choose between such functions when they give substantially different estimates of $E\{\mu\}$? By extension, in addition to the few functions tried, there must be many other functions which away from the origin will yield different estimates of $E\{\mu\}$ and make the data nearly equally likely. While there is no compelling reason to prefer one such function to another their estimates of $E\{\mu\}$ can be substantially different. In this and in similar cases the uncertainty about the estimate of $E\{\mu\}$ is in large part due to the uncertainty about which of the many alternative functions that could be chosen is the right one.

Table 10.3 A few model equations fitted by maximizing the NM likelihood function

$f_2(X_2)$	β_1[a]	β_2	β_3	$\beta_{Rolling}$	$\beta_{Mountainous}$	$\boldsymbol{b}$	Log-likelihood
1. $X_2^{\beta_2}$, Power, All data	0.968	0.800		1.877	2.914	54.74	−25,930
2. $X_2^{\beta_2}$, Power, AADT $< 3,300$	1.092	0.525		1.619	2.436	29.28	−20,623
3. $X_2^{\beta_2}(1+\beta_3 X_2)$, Power × Linear	0.968	0.784	0.143	1.878	2.917	54.96	−25,929
4. $X_2^{\beta_2}\left(1+\frac{\beta_3}{X_2}\right)$, Power × Hyperbolic	0.968	1.784	6.894	1.878	2.917	55.00	−25,929
5. $X_2^{\beta_2}e^{\beta_3 X_2}$, Hoerl	0.968	0.787	0.114	1.878	2.917	54.96	−25,929
6. $X_2(1+\beta_2 X_2+\beta_3 X_2^2)$, Polynomial	0.986	−1.49	1.157	1.793	2.777	53.95	−25,980
7. $\frac{1}{1+\beta_2 X_2^{\beta_3}}$, $\beta_2 < 0$, Sigmoid	0.968	1129	−0.80	1.877	2.914	54.73	−25,930

[a] In $X_1^{\beta_1}$

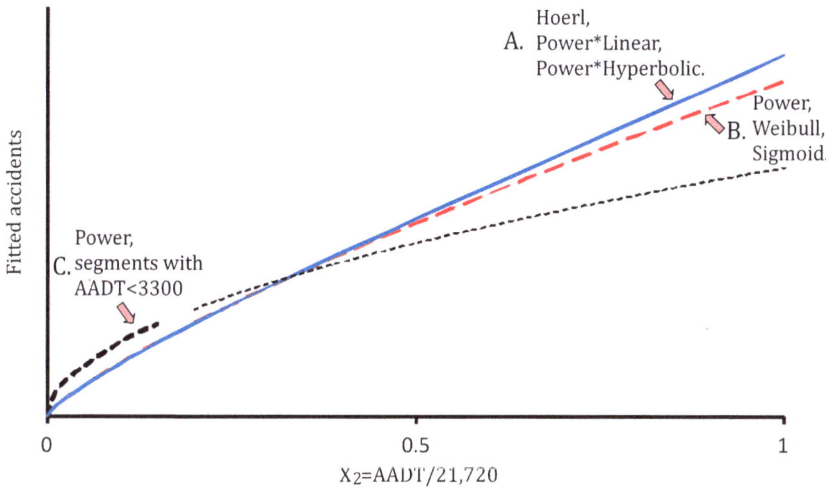

Fig. 10.11 Selected model equations and their graphs

This source of uncertainty is not usually considered when estimating the variance of $\hat{E}\{\mu\}$. The usual approach is assume that the variance of $\hat{E}\{\mu\}$ reflects the uncertainty surrounding the parameter estimates while the model equation used is assumed to be the correct one.[21] But, unless there is some preexisting theory, the chosen model equation is not known to be correct and in some cases, perhaps in most, there are many footloose functions, not one that dominates.

Different functions that fit to the same data almost equally well differ in their estimates of $E\{\mu\}$.

This then is yet another consideration which makes model equations footloose. In Chap. 9 it became evident that using the same data and model equation one could obtain substantially different estimates of $E\{\mu\}$ just by choosing one objective function rather than another. Here the footloseness is that of diverse functions that provide differing estimates of $E\{\mu\}$ having approximately the same likelihood. The upshot of the multifaceted footloseness is that the accuracy of the estimate of $E\{\mu\}$ depends not only on the uncertainty inherent in data and in parameter values, it also depends on the uncertainty about what mathematical expression should feature in the model equation and what objective functions ought to be used. The issue of estimate accuracy will be discussed in Chap. 11.

[21] See e.g., Wood (2005), Lord (2008), and Lord et al. (2010).

Curve C in Fig. 10.11 is the power model fitted to only those segments with average AADT < 3,300 (i.e., $X_2 < 0.15$). This curve was added in order to explain why the CURE plot remained recalcitrant when all data were used, no matter what function was tried. In the AADT range between 0 and 3,300 curve C fits the data well. None of the functions fitted to all the data could come as high because of the influence of data points in the high AADT region. Conversely, when the curve C fit is extrapolated into the high AADT region – the light dashes – the fit is not good. Thus, none of the single-equation models tried gave the curve the required shape in the entire AADT (X_2) range. It follows that in the case of the Colorado data set, and perhaps generally, it is not easy to find a relatively simple function that suits the data along its entire domain. For this reason, were one of these single-equation models chosen, it would act as a straightjacket to the SPF and cause bias-in-fit.[22]

10.7 Parameter Proliferation

There are, of course, many other functions that could be tried. Adding complexity and parameters might improve the fit. Doing so raises a sticky question: is the improvement in log-likelihood an indication of genuine progress or is it an artifact of parameter proliferation?

To explain, it helps to think of a parameter as a control knob which, as it is turned, changes the shape of the curve. Generally, the more knobs there are to turn the larger is the range of shapes a curve can take on and the better it can fit to the data. In the limit, if one used as many parameters as there are data points, all residuals would be zero; such a "model equation" would not capture any regularity hidden in the data, it would merely replicate what was observed.

The same issue arose earlier and in a more intuitive form in the context of nonparametric curve-fitting.[23] By choosing a narrow bandwidth one could make the fitted curve come closer to the data points while choosing a wider bandwidth was seen to dampen oscillations whereby the distance between the fitted curve and the data points increased. It was difficult to know what the best choice of bandwidth is. Similarly, in the present context, one can improve the fit by increasing the number of parameters but it is difficult to know where parameter inflation and "overfitting" begins.[24]

Modelers address the question formally using the Akaike or the Bayesian "Information Criteria," the acronyms of which are AIC and BIC. The better

[22] See Sect. 4.2.

[23] See Sect. 4.3.

[24] In SPF modeling when for some subset of units the sum of squared residuals is smaller than the sum of fitted values, the crash counts are less widely dispersed around the fitted value than what is consistent with the Poisson distribution; such as model is most likely "overfitted."

model is that for which the AIC or the BIC are smaller.[25] The AIC is 2(number of parameters in model) – 2ln(maximized likelihood). According to this criterion it is worth adding a parameter if doing so increases the natural logarithm of the maximized likelihood by more than 2.7. The BIC is 2ln(maximized likelihood) + (number of parameters in model)ln(number of data points). By this criterion it is worth adding a parameter if doing so increases the ln(maximized likelihood) by more than ln(number of data points)/2.

To illustrate, model Eq. (3) in Table 10.3 has one more parameter than model Eq. (1) and the addition of this parameter increased the log-likelihood from −25,930.17 to −25,929.72, i.e., by 0.45. The AIC requires an increase by at least 2.7 and the BIC by more than ln(5,323)/2 = 4.3. In this case one would go with the simpler model Eq. (1).

10.8 Options and Choices: Terrain Revisited

The question of how a predictor variable should be represented in a model equation is not as narrow as that of choosing one mathematical function out of many. To illustrate, in Sect. 9.6 the provisional decision there was to represent Terrain by two multipliers: β_{Rolling} and $\beta_{\text{Mountainous}}$ (with $\beta_{\text{Flat}} = 1$).[26] Another option for representing Terrain in the model equation is to fit three separate SPFs, one to each terrain. In addition, as noted during the EDA,[27] it is possible that the influence of the Terrain predictor variable on the $E\{\mu\}$ depends in some way on both Segment Length and on AADT. Therefore a third option would be to make β_{Terrain} into a function of these predictor variables. In short, there are many ways in which the influence of a predictor variable on the dependent variable can be accommodated in the SPF and it is up to the modeler to make the choice. To make an informed choice the ramifications of the various options need to be examined.

10.8.1 Fitting Separate SPFs

To fit a model to each terrain the data in Fig. 9.11 were separated into F, R, or M spreadsheets. In each such C-F spreadsheet the model equation is $\hat{E}\{\mu\} = \beta_0 X_1^{\beta_1} X_2^{\beta_2}(1 + \beta_3 X_2)$ and the $\sum|\text{residuals}|$ was minimized to yield the parameter estimates in Table 10.4.[28]

[25] The AIC is based on information theory, the amount of information lost when a given model is used to represent the process that generated the data. The BIC is based on comparing the posterior probability of the data of models the prior probability of which is assumed to be the same. The BIC tends to be the more stringent of the two criteria.

[26] The current estimates of β_{Rolling} and $\beta_{\text{Mountainous}}$ are in Fig. 10.6.

[27] See Sect. 3.6.

[28] To download these spreadsheets go to http://extras.springer.com/ and enter the ISBN of this book. Look in the "Spreadsheets" folder for "Chap. 10. Flat, Power x Line.xls or xlsx," "Chap. 10. Rolling, Power x Line.xls or xlsx" and "Chap. 10. Mountainous, Power x Line.xls or xlsx."

The solid curves in Fig. 10.12 are based on the parameters in Table 10.4 and the dashed curves on the estimates of β_{Rolling} and $\beta_{\text{Mountainous}}$ in Fig. 10.6. The common normalized segment length was 0.1 (2 miles). While for mountainous terrain the two curves are close, for Flat and Rolling terrains the two modeling options produce substantially different estimates of $E\{\mu\}$.

One of the differences between these modeling options is in the number of parameters used. The model $\hat{E}\{\mu\} = \beta_0 X_1^{\beta_1} X_2^{\beta_2}(1+\beta_3 X_2)\beta_{\text{Terrain}}$ with β_{Rolling} and $\beta_{\text{Mountainous}}$ ($\beta_{\text{Flat}} = 1$) requires 4 + 2 parameters. To fit $\hat{E}\{\mu\} = \beta_0 X_1^{\beta_1} X_2^{\beta_2}$ $(1+\beta_3 X_2)$ for each terrain 4×3 parameters are needed.

The merit of adding or of saving parameters can be quantified in light of the AIC and BIC discussed in the preceding section. The log-likelihood of the "Common model × multipliers" in Fig. 10.6 was −26,088. The sum of log-likelihoods in Table 10.4 is −25,959. Thus the addition six parameters increased log-likelihood

Table 10.4 Separate models for Terrain = F, R, and M

			X_1	X_2			
Terrain	Accidents	β_{scale}	β_1	β_2	β_3	ℓ	Log-likelihood
Flat	1,882	39.09	1.204	0.889	3.315	47.76	−4,773.7
Rolling	8,563	14.77	0.982	0.574	4.100	103.47	−13,442.4
Mountainous	11,273	39.48	1.070	0.689	3.273	36.86	−7,742.8
						Sum	−25,958.9

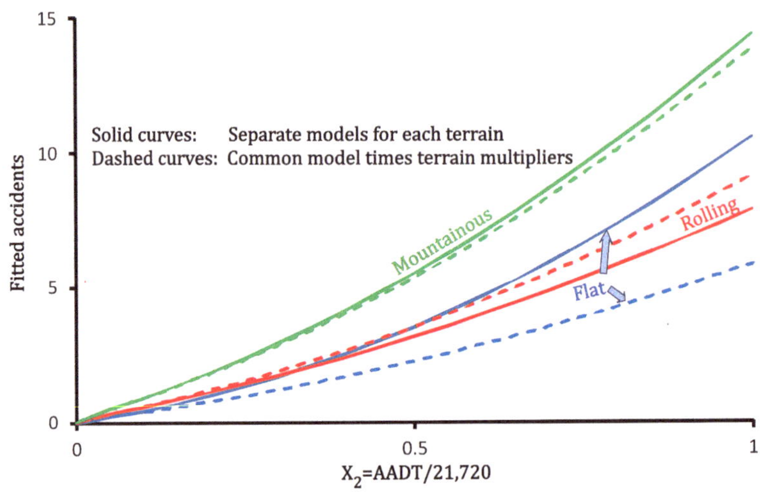

Fig. 10.12 Comparing "Common model × multiplier" versus "Separate models"

by 129 which is easily justified by both by both AIC and BIC.[29] On this score then one might prefer the "Separate models" option.

As noted earlier, the "Common model × multipliers" option rests on the assumption that β_1, β_2, and β_3 are the same for all three terrains. There is little reason to think this to be true.[30] Inasmuch as the estimates in Table 10.4 do not support the assumption, on this score too, the "Separate models" option is more attractive.

While the "Separate models" option has two strikes in its favor, the modeler cannot yet make the choice because the main question remains unanswered. The main question is not which option requires fewer parameters, neither which requires fewer unsupported assumptions, nor which makes the data more likely. When the focus is on applications[31] and the purpose is to produce good estimates of $E\{\mu\}$ the main question is which SPF will do so.[32] The quality of estimates question will be discussed in Chap. 11.

The CURE plots for mountainous terrain in Fig. 10.13 are based on parameters estimated by minimizing $\sum|\text{residuals}|$. The CURE plots for the rolling terrain are similar. Compared to the baseline CURE plots in Fig. 10.3 the use of separate models by terrain is a substantial improvement.

The parameter estimates in Table 10.4 (and the CURE plots in Fig. 10.13) are based on minimizing $\sum|\text{residuals}|$. For completeness and to facilitate comparison, the parameters for the mountainous terrain data were also estimated by maximizing the NM likelihood. Both sets are in Table 10.5 and the corresponding two curves are compared in Fig. 10.14. Several observations follow.

First, the choice of the objective function is seen to matter but the grounds on which the choice is to be made are not well defined. By habit and tradition road safety modelers tend to maximize likelihood. However, if the aim is to predict well then, as this case suggests, this may not always be the right choice. Those interested

[29] While both AIC and BIC use the maximized log-likelihood the values in the rightmost column of Table 10.4 and in Fig. 10.6 are based parameter estimates obtained by minimizing the sum of absolute residuals. Even so, increase of the log-likelihood of 129 is close to what would obtain if all parameters were estimated by maximizing likelihood.

[30] One might be inclined to conduct a statistical test of significance with the "all-parameters-are-the-same" assumption as the null hypothesis. This inclination is best resisted. The problem is that such a statistical test does not answer the question of interest; it does not say what the probability of the aforementioned assumption to be (approximately) true is. The statistical test only speaks about the probability of obtaining parameter estimate differences equal to or larger than those in Table 10.4 if the null hypothesis was true. If this probability is small (say, less than 0.05) then the "no difference" hypothesis "rejected"; otherwise it is "not rejected." If the "no difference" hypothesis is not rejected then, in spite of one's better judgment, one is stuck with the "no-difference" hypothesis even though the "yes-difference" hypotheses are more plausible It makes no common sense to proceed on the basis an un-rejected "no difference" assumption when a "yes-difference" assumption is better supported by the data. For discussion see, e.g., Edwards (1976, pp. 179, 180), Harlow et al. (1997), Hauer (1983).

[31] See Sect. 1.5.

[32] See Sect. 2.3.

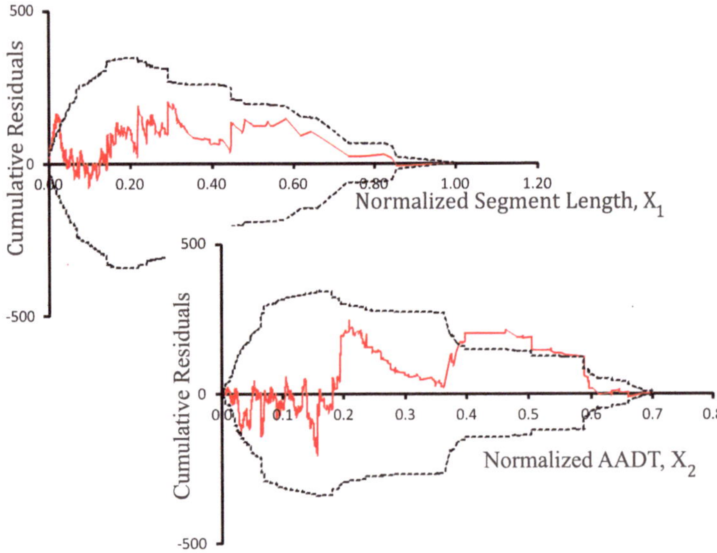

Fig. 10.13 CURE plots for mountainous terrain

Table 10.5 Comparing parameter estimates (Mountainous) for two objective functions

Objective function		X_1	X_2				
	β_{scale}	β_1	β_2	β_3	b	Log-likelihood	$\sum\lvert\text{residuals}\rvert$
Maximize likelihood	46.12	0.924	0.739	0.125	35.38	−7,674.0	6,197.9
Minimize $\sum\lvert\text{residuals}\rvert$	39.48	1.070	0.689	3.273	36.86	−7,742.8	5,955.5

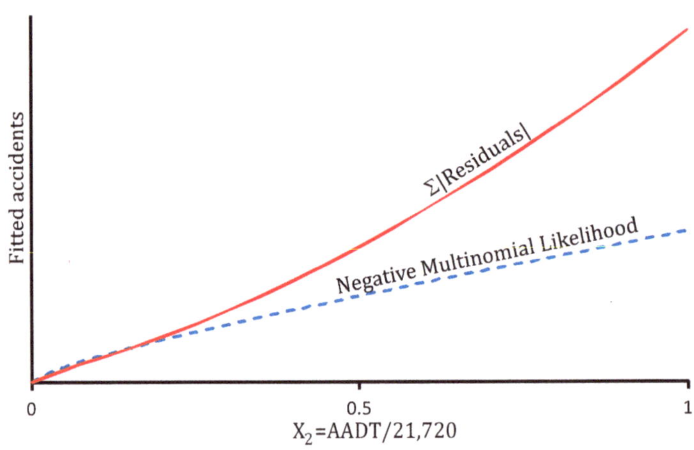

Fig. 10.14 Comparing curves obtained by two objective functions

in the properties of parameter estimates may well continue to maximize likelihood; those interested in producing good estimates of $E\{\mu\}$ should consider alternative objective functions.

Second, maximizing or minimizing an objective function was said to be the Archimedean lever which enables us to estimate parameters. There are several levers for the modeler to choose from and each places the hypersurface which the model equation represents in a different position within the same data cloud. This is another instance of the "footloseness" of model equations noted earlier in Sect. 9.7.

Third, that parameter estimates (such as those in Table 10.5) depend on what the modeler chooses to minimize or maximize may be found disconcerting. However, it is disconcerting mainly when the regression equation is used to predict by how much $E\{\mu\}$ will change when a predictor variable is changed by a specified amount. In this case one can rightly ask: "Why should the effect of a treatment or of a manipulation depend on the choice of the objective function?" As noted earlier,[33] such uses are to be questioned. It is not disconcerting when the model equation is used to estimate the average number of injury and fatal accidents for a population of rural two-lane road segments of a certain length, with a stated AADT, in a given year, and in a specified terrain of Colorado.

10.8.2 Making β_{Terrain} into a Function of Other Predictor Variables

Early on, during the initial EDA, it appeared that in rolling and mountainous terrain the association between $E\{\mu\}$ and the Terrain variable may depend on both Segment Length and AADT.[34] Attempting to capture this "interaction," instead of the constant β_{Terrain} in $\hat{E}\{\mu\} = \beta_0 X_1^{\beta_1} X_2^{\beta_2}(1+\beta_3 X_2)\beta_{\text{Terrain}}$, the function f_3 in Eq. (10.4) is used.

$$f_3(\text{Terrain}, X_1, X_2, \boldsymbol{\beta}) = \begin{cases} 1 & \text{if Flat} \\ \beta_{\text{Rolling}}(1+\beta_4 X_1 + \beta_5 X_2) & \text{if Rolling} \\ \beta_{\text{Mountainous}}(1+\beta_6 X_1 + \beta_7 X_2) & \text{if Mountainous} \end{cases} \tag{10.4}$$

The corresponding parameter estimates obtained by maximizing the NM likelihood and by minimizing the $\sum|\text{residuals}|$ are in Fig. 10.15.[35]

The purpose of this section was to show that there are several distinct ways in which to include a predictor variable in the model equation. The predictor variable here was Terrain and it was alternatively represented by:

---

[33] See Sects. 6.7 and 10.2.

[34] See Sect. 3.6, Fig. 3.16.

[35] To download this spreadsheet go to http://extras.springer.com/ and enter the ISBN of this book. Look in the "Spreadsheets" folder for "Chap. 10. Terrain as a function of X1 and X2.xls or xlsx."

Objective function	β_{scale}	X_1, Length/19.74	X_2, AADT/21,720		X_3, Terrain						b
		β_1	β_2	β_3	$\beta_{rolling}$	β_4	β_5	$\beta_{mountainous}$	β_6	β_7	
NM Likelihood	12.51	0.997	0.616	-0.247	1.479	0.035	2.829	2.026	-0.077	4.517	55.205
Σ|residuals|	14.17	1.044	0.666	3.640	1.627	-0.014	-0.114	2.385	-0.024	-0.021	57.301

Objective function	Log Likelihood	Σ|residuals|
NM Likelihood	-25834.7	12052.5
Σ|residuals|	-26031.6	11882.6

Fig. 10.15 Terrain as a function of Segment Length and AADT

(a) Two constants β_{Rolling} and $\beta_{\text{Mountainous}}$.
(b) Two functions $\beta_{\text{Rolling}}(1+\beta_4 X_1+\beta_5 X_2)$ and $\beta_{\text{Mountainous}}(1+\beta_6 X_1+\beta_7 X_2)$.
(c) Three separate model equations, one for each terrain.

Option (a) required six parameters, option (b) 10, and option (c) 12. As expected, the larger was the number of parameters the larger was the log-likelihood and the lesser was the $\sum|\text{residuals}|$. The question of which option makes for more accurate estimates and predictions awaits examination in Chap. 11.

Option (b) is a departure from the multiplicative model in which each factor is a function of only one predictor variable. To allow for the possibility that the association between Terrain and $E\{\mu\}$ may not be the same on short and long segments and/or that it may depend on whether traffic is light or heavy, the functions in option (b) depends not only on Terrain (through β_{Rolling} and $\beta_{\text{Mountainous}}$) but also on Segment Length (through X_1) and on AADT (through X_2).

In road safety interactions of this kind are common. One may expect that the influence of the Lane Width as predictor variable may depend on Shoulder Width and on whether it is Paved or Unpaved; one can assume that the contribution to $E\{\mu\}$ of the Radius of Horizontal Curvature is influenced by preceding Frequency of Curves and Tangent Length; one can anticipate that the effect of Left-Turn Protection at a traffic signal is a function of the Number of Opposing Lanes and of Approach Speed, etc. Because in road safety such interactions are of essence, when the modeler chooses the structure of the model equation thought must be given to their accommodation. This topic is discussed next.

10.9 Interaction

"An interaction effect is said to exist when the effect of the independent variable on the dependent variable differs depending on the value of a third variable, called the moderator variable." Jaccard and Turrisi (2003). Is the "Terrain" variable subject to interaction? Is AADT a moderator variable for Terrain?

When used as a predictor (independent) variable "Terrain" is a proxy for grade, sight distance, curvature, lane and shoulder width, sideslope, etc., all variables that are associated with Terrain and thought to affect the probability of accident occurrence and accident severity. The question is whether the influence on $E\{\mu\}$ of a variable such as Lane and Shoulder Width (and thereby of Terrain) varies with, say, AADT. Empirical evidence suggests that the answer is affirmative.[36] The same is likely true for grade, sight distance and other safety-related variables associated with Terrain. If so, AADT is likely to be a moderator variable for Terrain and the modeler ought to allow for this interaction when shaping the model equation.

The narrow question is whether the three options for incorporating Terrain into the model equation discussed earlier can capture the AADT–Terrain interaction. Option (a) clearly does not do so; the two multiplier-constants β_{Rolling} and $\beta_{\text{Mountainous}}$ are the same irrespective of what AADT is. Option (b) is a deliberate and seemingly sensible attempt to capture that interaction; whether it succeeds will be examined below. Option (c) appears capable of expressing the interaction. The three solid curves in Fig. 10.12 each depend on AADT in a somewhat different manner and therefore their ratio will be a function of AADT. Whether a function obtained in this manner is more than a arbitrary by-product of curve-fitting will be discussed.

The broader question is how to provide for interaction when the modeler is shaping the $f()$ in $E\{\mu\} = f(X_1, X_2, \ldots, \beta_0, \beta_1, \ldots)$. As noted earlier, not much has been written on the general task of how to come up with a good $f()$. Guidance on the sub-task of capturing interactions in multivariable regression models is even sparser.

In road safety modeling there are two main traditions for obtaining the algebraic expressions which describe how $E\{\mu\}$ depends on predictor variables. The econometric tradition is to examine "elasticity,"[37] i.e., the percent change in $E\{\mu\}$ due to 1 % change in X_i. The road safety management tradition is to examine the Crash Modification Function (CMF), i.e., the ratio $E\{\mu|X_{i,a}\}/E\{\mu|X_{i,b}\}$. Both traditions invoke the "ceteris paribus" assumption; namely they examine the change in $E\{\mu\}$ when only the level of X_i changes from "b" (connoting "before") to "a" (connoting "after") while all else remains unchanged.

Section 10.3 describes the sequence of (unconvincing) arguments by which the general model equation $E\{\mu\} = f(X_1, X_2, \ldots, \beta_0, \beta_1, \ldots)$ was whittled down to the simplified $E\{\mu\} = e^{\Sigma_{\forall i}\beta_i X_i}$ which is now widely used in road safety modeling. In this book, so far, the less restrictive $E\{\mu\} = \beta_0 f_1(X_1, \boldsymbol{\beta}_1) f_2(X_2, \boldsymbol{\beta}_2) \ldots$ form was used. However, even this less restrictive family of model equations still retains two basic traits: (a) it is a product of factors and (b) each factor is a function on one variable only.

---

[36] Thus, e.g., having a 10′ lane on a rural two-lane road instead of a 12′ lane is expected to increase the number of accidents by a factor of 1.3 when AADT = 2,000 and by a factor of 1.11 when AADT = 1,000 (AASHTO 2010, Figs. 13.1 and 13.5).

[37] For the general model equation $E\{\mu\} = f(X_1, X_2, \ldots, \beta_0, \beta_1, \ldots)$, the elasticity of function f with respect to variable X_i is defined as $\varepsilon_{f,X_i} \equiv \frac{\partial f}{\partial X_i}\frac{X_i}{f}$.

Contrary to common belief,[38] model equations of this family cannot possibly accommodate interaction.[39] Whatever the functions $f_1, f_2, \ldots$, be they simple (as in $e^{\sum_{\forall i}\beta_i X_i}$) or more complicated, the corresponding expression for elasticity is always a function of one variable only and the corresponding CMF is never a function of moderator variables.

A model equation that is a product of single variable factors cannot accommodate interaction and cannot be a source of CMFs that are a function of other variables.

This negative conclusion is important. It should diminish the waste of time and resources in the pursuit of an illusion. After all, it is clear that Shoulder Width is a moderator for Lane Width, Tangent Length for Degree of Curve, Sight Distance for Pavement Friction, etc. Because interaction is omnipresent in the real world of road safety, continued reliance on model equations that are products of single-variable factors appears to be folly.

The positive aspect of the same conclusion should motivate the pursuit of means by which interaction can be incorporated into model equations.[40] The first step is already clear: for a factor in a multiplicative model to accommodate interaction, it must be a function of the relevant moderator variables. The next step is to ask what form this function should take.

The Jaccard and Turrisi (2003) book from which the opening quote was taken is about interaction in additive linear regression models. The authors show, for example, that if $Y = \beta_1 X_1 + \beta_2 X_2$, and if X_2 is a moderator variable that affects β_1 through $\beta_1 = \alpha_0 + \alpha_1 X_2$, then the model equation changes into $Y = \alpha_0 X_1 + \beta_2 X_2 + \alpha_1 X_1 X_2$. That is, β_1 changes to α_0 and the interaction between X_1 and X_2 is captured by adding the product term $\alpha_1 X_1 X_2$ to the right hand side.[41]

Road safety models are multiplicative (not additive) and functions representing predictor variables are not of the simple $\beta_i X_i$ form. The question is by what algebraic expression to represent interaction in such models. Consider, for example, the multiplicative model $Y = \beta_0 \times X_1^{\beta_1} \times X_2^{\beta_2}$ in which X_2 is a moderator variable that affects β_1 through $\beta_1 = \alpha_0 + \alpha_1 X_2$. As shown in Appendix M, to accommodate this interaction the model has to be $Y = \beta_0 \times X_1^{\alpha_0} \times X_2^{\beta_2} \times X_1^{\alpha_1 X_2}$.

---

[38] Thus, e.g., Chen and Persaud (2014, p. 132) assert that in their model equation $E\{\mu\} = \beta_0$ $\text{AADT}^{\beta_1} e^{\beta_2 X_2} e^{\beta_3 X_3} \ldots$ the factors $e^{\beta_2 X_2}$, $e^{\beta_3 X_3}$, … "are technically CM-Functions" for the X_2, $X_3, \ldots$ Since theirs is a multiplicative model equation in which each factor is a function of only one predictor variable, as shown in Appendix L, the assertion is open to question.

[39] See Appendix L.

[40] Whether CMFs obtained from such model equations can be trusted is a different question as discussed in Sect. 6.7.

[41] For detail see Appendix M.

Some general conclusions follow. First, if the influence of predictor variable X_i on $E\{\mu\}$ depends on the value of moderator variable X_m then, for the model equation to accommodate this interaction one has to add a new "interaction factor" which is a function of both X_i and X_m. Second, the form of the interaction factor is not always the same; in one circumstance the term $\alpha_1 X_1 X_2$ was added to the model equation, in another circumstance $X_1^{\alpha_1 X_2}$ multiplied the model equation. The interaction terms depends on the function by which X_i alone was assumed to influence $E\{\mu\}$ and the manner in which X_m is assumed to influence this function. Third, the addition to the model equation of the interaction factor will change the parameters in the function by which X_i alone was assumed to influence $E\{\mu\}$.

The above procedure for structuring model equations that can accommodate interactions has attained a degree of formalization under the rubric of Hierarchical (or alternatively, Multilevel) modeling.[42] At the first level is the original $E\{\mu\} = \beta_0 f_1(X_1, \boldsymbol{\beta}_1) f_2(X_2, \boldsymbol{\beta}_2) \ldots$ – the product of single-variable factors. The equations by which the parameters vectors $\boldsymbol{\beta}_1$, $\boldsymbol{\beta}_2, \ldots$ depend on the predictor variables $X_1, X_2, \ldots$ constitute the second level of the hierarchy. Substituting for the $\boldsymbol{\beta}$s in the first level the equations from the second level makes some factors into a function of several variables and thereby accounts for interaction.

Option (b) in Sect. 10.8.2 for the addition of the Terrain variable is an illustration. Here the β_{Rolling} and $\beta_{\text{Mountainous}}$ were replaced by $\beta_{\text{Rolling}}(1 + \beta_4 X_1 + \beta_5 X_2)$ and by $\beta_{\text{Mountainous}}(1 + \beta_6 X_1 + \beta_7 X_2)$. If the linear relationships well represent the influence of the moderator variables X_1 and X_2 then, as shown in Appendix M, this option is satisfactory.

The third option mentioned there (Option c) was to extract interaction from separately estimated models. Thus, for example, using the parameters from Table 10.4 in $\hat{E}\{\mu\} = \beta_0 X_1^{\beta_1} X_2^{\beta_2}(1 + \beta_3 X_2)$ the estimate of the $E\{\mu_{\text{Mountainous}}\}/E\{\mu_{\text{Rolling}}\}$ is $2.67 X_1^{0.088} X_2^{0.115} \frac{1+3.273X_2}{1+4.100X_2}$. This function is an automatic by-product of the model equation chosen earlier. Had a different model equation been chosen or different objective function optimized, a different result would obtain. Because the chosen model equation is so footloose, the function representing the $E\{\mu_{\text{Mountainous}}\}/E\{\mu_{\text{Rolling}}\}$ ratio is also uncertain. The main weak point of option (c) is that it requires neither insight into nor deliberation about what the moderator variables are and how they might be expressed in the model equation. Before such deus-ex-machina results can be trusted, more research is required.[43]

---

[42] See, e.g., Chen and Persaud (2014, Sect. 4).

[43] In this specific case it produces suspicious results for low-volume short road segments. Thus, e.g., for a 0.2 miles long segment with AADT = 100 it predicts fewer accident in mountainous than in rolling terrain.

10.10 Summary

The question was how to shape the $f()$ in $E\{\mu\} = f(X_1, X_2, \ldots, \beta_0, \beta_1, \ldots)$; this was said to be the holy grail of modeling, the promise of good predictions, and the condition for cause-effect interpretations. That on so central an issue there is so little guidance is vexing. One can lean mostly on the broken reeds of habit, convention (much of it groundless), goodness-of-fit, and parameter parsimony.

To show how elusive $f()$ can be a story was concocted: A researcher is given good data to determine how Y depends on X_1 and X_2. Based on an EDA the researcher fits a model equation which does an excellent job of predicting what Y would be, on the average, when the magnitudes of X_1 and X_2 are given. The same model and therefore also the researcher fail badly when predicting what change in Y is caused by a change in X_1 or X_2. This is so because, based on an EDA, the researcher assumed that $f()$ is $\beta_0 + \beta_1 X_1 + \beta_2 X_2$ when, in fact it was $\left(X_1^{\beta_1} + X_2^{\beta_2}\right)^{\beta_3}$. With plentiful and good data the regression did not find the Pythagoras theorem. The main moral is that you do not have to have the right $f()$ in order to estimate Y, but that having the right $f()$ is essential if you want to predict the effect of manipulation, intervention, and change.

Why are so many modelers content to just state that $E\{\mu\} = e^{\sum \beta_i X_i}$ is the model equation? I found no good answer to this question. The manner in which the general expression $E\{\mu\} = f(X_1, X_2, \ldots, \beta_0, \beta_1, \ldots)$ was whittled down to a product of simple identical functions is replete with questionable simplifications. The resulting simple log-linear structure is not likely to well represent the complexity of the process by which variables combine to generate accidents. Removing some of the inessential simplifications the more flexible structure $E\{\mu\} = \beta_0 \times f_1(X_1, \boldsymbol{\beta}_1) \times f_2(X_2, \boldsymbol{\beta}_2) \times \cdots \times$ (interaction terms) was adopted for further modeling. While this $f()$ is still a product of factors, the functions $f_1, f_2, \ldots$ can be complex and differ from one variable to another.

At the end of Chap. 9, with $f()$ the simple power function and Segment Length, AADT and Terrain the predictor variables, the CURE plots were very poor. The question was how to improve the inadequate model equation. Two remedies were tried. One was to apply "bump functions" to some region of the prediction variable, the other was to substitute alternative functions for the power model. These modifications and explorations were simple to implement in the C-F spreadsheet and improved the CURE plots substantially.

To make the transition from a graph that shows some data-based regularity to a mathematical function that can represent it, one has to know what functions look like. A spreadsheet tool was created for this purpose and its uses were described. Equipped with the ability to visualize the shape of functions and how they respond to change in parameters, several such functions were fit to the Colorado data. Function that in their algebraic form look very different were found to be nearly congruent near the origin but away from the origin they could give very different estimates of $E\{\mu\}$. None of the functions tried fitted the data well throughout its entire domain; every such function, if used, would cause bias-in fit in a part of the domain.

By increasing the complexity of function and the number of their parameters one can increase likelihood. The question is how much of an increase in likelihood justifies the addition of a parameter. Guidance was provided in terms of the Akaike and Bayesian Information Criteria.

To this point the question of how to shape *f*() was construed narrowly as that of choosing the algebraic expression for the functions f_1, $f_2, \ldots$ in $E\{\mu\} = \beta_0 \times f_1(X_1, \boldsymbol{\beta}_1) \times f_2(X_2, \boldsymbol{\beta}_2) \times \cdots \times$ (interaction terms). But the question of how to represent a variable in the model equation is broader than that. Thus, for example, Terrain could be represented either by two multiplicative constants, or by two functions of Segment Length and AADT, or by fitting separate models to data from each terrain. These options were examined in detail and their strengths and weaknesses discussed. It became obvious is that the fit of the model and the estimates and predictions of $E\{\mu\}$ depend crucially on what the modeler chooses to minimize of maximize. With the data at hand, minimizing the sum of absolute residuals seems to outperform the maximization of likelihood.

Another key element of shaping *f*() is enabling it to represent "interaction" when present. Interaction exists when effect on $E\{\mu\}$ of a predictor variable depends on the value of another, a "moderator" variable. In road safety interaction is omnipresent; the effect of Lane Width depends on Shoulder Width, of Horizontal Curvature on Tangent Length, etc. As it turns out, a model equation that is a product of single variable factors cannot accommodate interaction. When interaction is present the corresponding factor must be a function of the moderator variables. How this is to be done was explained.

The several improvements to the *f*() of the Colorado SPF led to CURE plots that are respectable. However, some of the modeling choices made were based on secondary considerations. The primary consideration, that of the estimate accuracy, is the subject of the next chapter.

References

AASHTO (The American Association of State Highway and Transportation Officials) (2010) Highway safety manual, 1st edn

Chen Y, Persaud B (2014) Methodology to develop crash modification functions for road safety treatments with fully specified and hierarchical models. Accid Anal Prev 70:131–139

Chiou Y-C, Fu C (2013) Modeling crash frequency and severity using multinomial-generalized Poisson model with error components. Accid Anal Prev 50:73–82

Edwards AWF (1976) Likelihood. Cambridge University Press, Cambridge

Harlow LL, Mulaik SA, Steiger JH (1997) What if there were no significance tests? Lawrence Erlbaum, London

Hauer E (1983) Reflections on methods of statistical inference in research on the effect of safety countermeasures. Accid Anal Prev 15(4):275–285

Hauer E (2004) Statistical safety modelling, Transportation research record 1897. National Academies Press, Washington, DC, pp 81–87

Hoerl AE (1950) Fitting curves to data. In: Perry JH (ed) Chemical engineer's handbook. McGraw-Hill, New York (Chapter 20)

Jaccard J, Turrisi R (2003) Interaction effects in multiple regression, 2nd edn. Sage University papers on quantitative applications in the social sciences, 07-072. Sage, Thousand Oaks

Lau LJ (1986) Functional forms in econometric model building. In: Griliches Z, Intriligator MD (eds) Handbook of econometrics, vol III. North-Holland, Amsterdam

Lord D, Bonneson JA (2007) Development of accident modification factors for rural frontage road segments in Texas. Transport Res Rec 2023:20–27

Lord D (2008) Methodology for estimating the variance and confidence intervals for the estimate of the product of baseline models and AMFs. Accid Anal Prev 40:1013–1017

Lord D, Kuo P-F, Geedipally SR (2010) Comparison of application of product of baseline models and accident-modification factors and models with covariates. Transport Res Rec 2147:113–122

Poch M, Mannering F (1996) Negative binomial analysis of intersection-accident frequencies. J Transport Eng 122:105–113

Shankar V, Mannering F, Barfield W (1995) Effect of roadway geometrics and environmental factors on rural freeway accident frequencies. Accid Anal Prev 27(3):371–389

Theil H (1971) Principles of econometrics. Wiley, New York

Wood GR (2005) Confidence and prediction intervals for generalized linear accident models. Accid Anal Prev 37:267–273

Accuracies

11

Abstract
While estimate accuracy is the touchstone for many modeling choices and decisions, its determination is difficult. The difficulty is not technical but conceptual. In this chapter the accuracy of estimates is determined by simulation. The simulation code has been used to examine several issues: how estimate accuracy is affected by the addition of the Terrain variable, how the variance of AADT estimates influences estimate accuracies, and what is the effect on accuracy of using fewer years of data.

11.1 Considerations

A good SPF is one that produces accurate estimates of $E\{\mu\}$ and small $\sigma\{\mu\}$'s. The matter of $\hat{E}\{\mu\}$ accuracy arose earlier and was seen to be the key to several modeling decisions. Already in Chap. 2[1] it was recognized that when $\hat{E}\{\mu\}$ helps to estimate the μ of a unit, the $V\{\hat{E}\{\mu\}\}$ is one of two elements determining the $V\{\hat{\mu}\}$. Later, in Chap. 9, when discussion revolved around the question of whether a variable should be added to the model equation, the decision could not be made without knowing by how much doing so will increase the $V\{\hat{E}\{\mu\}\}$. In Chap. 10, where the question was how to choose the model equation and whether increasing the number of required parameters is worthwhile, the answer again required knowing how adding parameters affects $V\{\hat{E}\{\mu\}\}$. The discussion of "accuracies" was postponed till this chapter, till when it became clear how the estimates which the SPF produces depend on the many modeling choices, decisions, and assumptions.

[1] See (2.2).

E. Hauer, *The Art of Regression Modeling in Road Safety*,
DOI 10.1007/978-3-319-12529-9_11

The inaccuracy of $\hat{E}\{\mu\}$ is usually attributed entirely to the uncertainty about the parameters of the model equation[2] and the latter is assumed to be solely due to the "statistical inaccuracy"[3] – that inaccuracy which stems mainly from the randomness of accident counts. There are, however, other important sources of uncertainty about $\hat{E}\{\mu\}$; those due to the lack of clarity about what objective function to use, those due to the footlessness of the functions selected to be the model equation, and those due to the absence of safety-related variables from the model equation.[4] These causes of "modeling inaccuracy," while present, are difficult to account for in some quantitative way. The reason is that one cannot meaningfully speak of the probability of an objective function being right, of the chance that some functional form of the model equation is the correct one, or of the bias due to the absence from the model equation of some safety-related variable. The upshot is that the estimate of $V\{\hat{E}\{\mu\}\}$ is a conditional one; it would be appropriate if the objective function was clearly the right one, if the model equation was truly behind the process that generated accidents counts, if there were no variables absent from the model equation, etc. As none of this is true, this conditional and partial estimate of $V\{\hat{E}\{\mu\}\}$ is too small, perhaps very much so. This is a major shortcoming. The most that can be done about this blemish at present is to declare it.[5]

One would like to make $V\{\hat{E}\{\mu\}\}$ into a touchstone that helps to make modeling decisions. Thus, for example, one would like to determine which of two alternative model equations makes $V\{\hat{E}\{\mu\}\}$ smaller. The problem is that the magnitude of $V\{\hat{E}\{\mu\}\}$ can be determined only when a specific model equation is assumed to be the right one and therefore the choice between alternatives remains indeterminate.

Only the conditional and incomplete $V\{\hat{E}\{\mu\}\}$ can be estimated and this can be done in several ways. Since the source of the $V\{\hat{E}\{\mu\}\}$ is assumed to be only the inaccuracy of parameter estimates, when least squares are used to estimate the parameters then its estimate comes from the variance-covariance matrix of the predictor variables. When likelihood is maximized the inverse of the Fisher Information Matrix is used for this purpose.[6] In what follows the $V\{\hat{E}\{\mu\}\}$ will be determined by simulation. The advantages of simulation are several. First, in a simulation one can represent all sources of statistical inaccuracy, including the "error in variables." Thus, for example, as will be shown, it is simple to account for the inaccuracy inherent in the estimates of AADT. Second, one is not limited

---

[2] See e.g., Wood (2005).

[3] See Sect. 6.6.1.

[4] These were called "modeling uncertainty" and discussed in Sect. 6.6.2.

[5] Mathematical deductions and the resulting expressions tend to impress by their rigor. However, when the premises of these deductions are assumptions, the results are conditional; they apply to the extent that the assumptions approximate reality. This is why assumptions made have to be stated explicitly and why they deserve the modeler's attention.

[6] See Sects. 6.6.1 and 8.2.

to parameter estimation by least squares or by maximum likelihood; one can estimate the accuracy of $\hat{E}\{\mu\}$ no matter what the objective function is. In this chapter $V\{\hat{E}\{\mu\}\}$ will be estimated when the sum of absolute residuals is minimized. Third, the simulation approach is explicit; it is clear what is being assumed and what the sources of uncertainty are. There is no need to rely on matrix inversion or on second order partial derivatives, constructs that are not transparent to intuition. As will be shown, the simulation code is organically coupled with the C-F spreadsheet. The VBA[7] code generates the data which the Solver then uses to find parameters and to compute fitted values. The main disadvantage of simulation is that, at present, the VBA code requires adaptation to specific circumstances and therefore, to be of use, some facility with VBA programming is a prerequisite.

11.2 The Simulation Idea

The idea of determining accuracies by simulation is best explained in the context of an illustration. To begin simply, suppose that the model equation is $\hat{E}\{\mu\} = \beta_0 X_1{}^{\beta_1} X_2{}^{\beta_2}$, that the data are those in Fig. 9.11, and that parameters are estimated by minimizing $\sum$|residuals|. These elements are shown in Fig. 11.1. The fitted values in that figure are the yearly estimates of $E\{\mu\}$. The question is what their standard error is.

Now the essence of the simulation idea can be described. Consider a population of road segments[8] with the same X_1 and X_2 as Segment 1. Suppose that the true sum of $E\{\mu\}$'s in this imagined population was indeed 0.17 accidents in 1986–1998 and that the true ℓ was 41.70. If so, the variance of $\sum\mu$ in this imagined population is $V\{\sum\mu\} = (\sum E\{\mu\})^2/(\ell X_1) = 0.17^2/(41.7 \times 0.076)$. These $\sum E\{\mu\}$ and $V\{\sum\mu\}$ define a Gamma distribution of $\sum\mu$'s. In the simulation the $\sum\mu$ for this road segment will be randomly generated from that Gamma distribution. The μ in year y will then be computed as $(\sum\mu) \times$ (Fitted value in year y)/(Sum of fitted values). Using the yearly μ's the number of accidents in each year will be generated from a Poisson distribution. The same will be done for all 5,323 segments. Using the new $13 \times 5{,}323$ matrix of newly generated accident counts, the parameters of the model equation will be reestimated and new fitted values obtained. The entire process is then repeated several times. Each such simulation "run" yields a new set of parameter estimates and fitted values. Using these, as will be shown, the variances of interest can be estimated.

The paragraph above presents the basic idea. The point of departure is the fitted values which are taken to be the true $E\{\mu\}$'s of an imagined populations. The sum of these, $\sum E\{\mu\}$, jointly with the parameter ℓ, yields the $V\{\sum\mu\}$. Next, a $\sum\mu$ is randomly generated from the Gamma distribution the moments of which are

[7] Visual basic for applications.

[8] See Sect. 1.2.

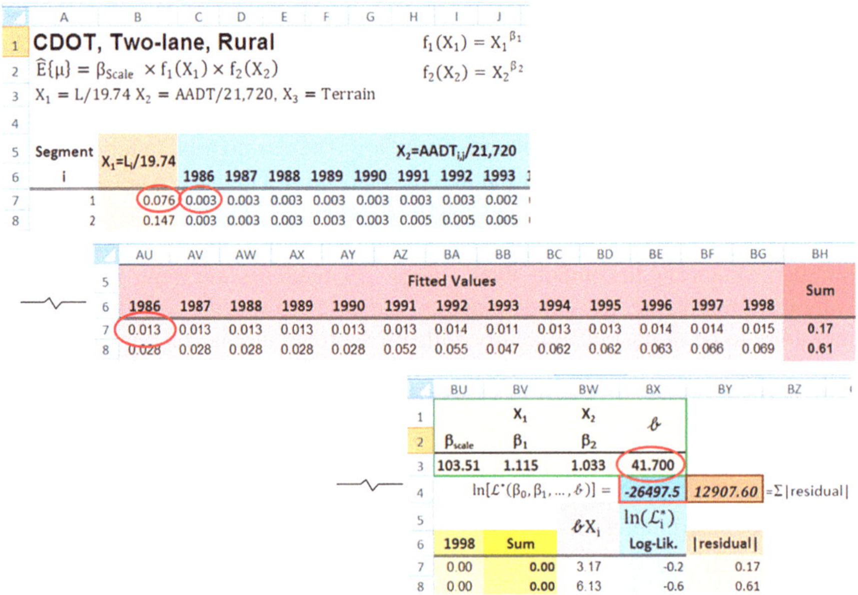

Fig. 11.1 Fitted values and parameter estimates

$\sum E\{\mu\}$ and $V\{\sum\mu\}$. The μ's for all years are then prorated. Using these yearly μ's, accident counts are randomly generated from the Poisson distribution. Based on these new accident counts Solver obtains new parameter estimates and new fitted values. Repeating such simulation runs several times, sample variances of both parameters and fitted values interest can be computed. How this is done is shown next.

11.3 The Idea Executed

Here the simulation idea will be implemented using a C-F spreadsheet with the Colorado data and $\hat{E}\{\mu\} = \beta_0 X_1^{\beta_1} X_2^{\beta_2}$ as the model equation.[9] On the spreadsheet there is a "command button" which activates the simulation code. To begin, the spreadsheet will be used to determine the standard errors of $\hat{E}\{\mu\}$ and of the parameter estimates. Next the effect on these standard errors of adding the Terrain variable will be examined. Following that, the contribution to the standard errors of inaccuracies in AADT will be quantified. Finally some study design issues will be explored.

[9] To download this spreadsheet go to http://extras.springer.com/ and enter the ISBN of this book. Look in the "Spreadsheets" folder for "Chap. 11. Simulation with $X1$ and $X2$.xls or xlsm."

11.3.1 Determining Standard Errors

The simulation was run ten times. The estimates of $\sum E\{\mu\}$ produced by the simulation for the first five road segments are in columns G to P of Fig. 11.2. Thus, for example, for Segment 1 for which the true $\sum E\{\mu\}$ was taken to be 0.17 accidents in 13 years, the simulation runs produced $\sum E\{\mu\}$ estimates of 0.14, 0.16, …, 0.17. For this segment the square root of the average squared difference around the 0.17, the standard error, is $\hat{\sigma}\left\{\Sigma\hat{E}\left\{\mu\right\}\right\} = 0.013$ accidents in 13 years; for Segment 2 it is 0.031, and so on.

The relationship between the true $\sum E\{\mu\}$'s in column F and the standard errors in column Q for all 5,323 segments is depicted in Fig. 11.3. The solid line is a fitted Hoerl curve[10]. The limitations noted in Sect. 11.1 apply. Namely, that the ordinate would be a valid estimate of the standard error of the estimate of $\sum E\{\mu\}$ if $\hat{E}\left\{\mu\right\} = \beta_0 X_1^{\beta_1} X_2^{\beta_2}$ was the right model equation, if no safety-related variables

	E	F	G	H	I	J	K	L	M	N	O	P	Q
3	Segment	TRUE	Simulation runs										$\hat{\sigma}\{\Sigma\hat{E}\{\mu\}\}$
4		$\Sigma E\{\mu\}$	1	2	3	4	5	6	7	8	9	10	
5	1	**0.17**	0.14	0.16	0.18	0.17	0.17	0.16	0.16	0.18	0.16	0.17	0.013
6	2	**0.62**	0.56	0.59	0.67	0.61	0.63	0.58	0.59	0.64	0.60	0.63	0.031
7	3	**0.29**	0.25	0.28	0.30	0.28	0.29	0.27	0.27	0.30	0.27	0.29	0.020
8	4	**0.07**	0.05	0.06	0.06	0.06	0.06	0.06	0.06	0.06	0.05	0.06	0.009
9	5	**0.85**	0.81	0.83	0.95	0.87	0.89	0.82	0.82	0.91	0.85	0.88	0.044

Fig. 11.2 Simulated estimates of $\sum E\{\mu\}$

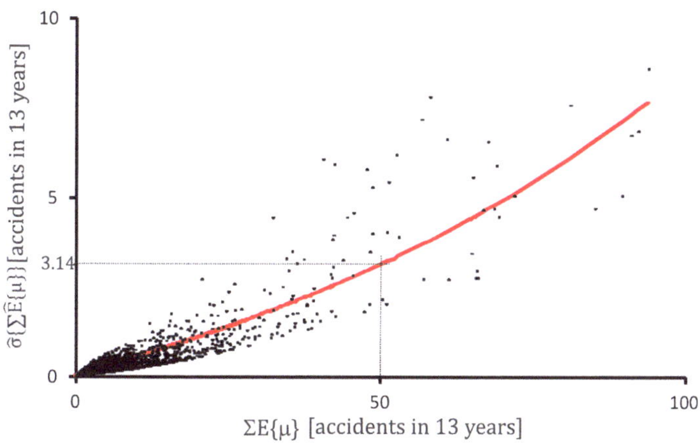

Fig. 11.3 The standard deviation of the estimate of $E\{\mu\}$

[10] $\hat{\sigma}\left\{\Sigma\hat{E}\left\{\mu\right\}\right\} = 0.0685(\Sigma E\{\mu\})^{0.878}\exp(0.0078\Sigma E\{\mu\})$.

were missing from it, and if minimizing $\sum$|residuals| was the correct objective for parameter estimation.

Thus, for example, for a population of segments expected to have 50 accidents during the 13 years period 1986–1998 the standard error of the estimate of $\sum E\{\mu\}$ is ± 3.14. Expressed as a yearly average, for a population of segments with $50/13 = 3.85$ accidents/year the standard error is $\pm 3.14/13 = \pm 0.24$ accidents/year.

Each simulation run produced also a set of parameter estimates from which their standard error was computed. The statistical accuracies[11] of the estimates of β_1 and β_2 for this model equation are listed in the usual manner[12] in the first row of Table 11.1.

11.3.2 How Accuracy Is Affected by the Addition of the Terrain Variable

In Chap. 9, when discussing the question of when to add a variable, the issue of whether the "sufficient condition" is met was left unresolved. To meet the sufficient condition one had to show that by adding a variable the bias-in-use was reduced by more than the standard error of $\hat{E}\{\mu\}$ was increased. But the means to estimate this standard error were not available. The change in the standard error of $\hat{E}\{\mu\}$ due to the addition of the Terrain variable can now be determined.

One option for adding Terrain into the model equation was[13] to multiply $\hat{E}\{\mu\} = \beta_0 X_1{}^{\beta_1} X_2{}^{\beta_2}$ by β_{Terrain}. Simulation showed that the addition of this variable left the accuracy with which the $E\{\mu\}$'s are estimated virtually unchanged.[14] It follows that, since adding the terrain variable reduces bias-in-use while $V\{\hat{E}\{\mu\}\}$ is hardly increased, the sufficient condition for adding Terrain is met.[15] The statistical uncertainty surrounding the parameter estimates for the "with-Terrain" model as determined by simulation is shown in the second row of Table 11.1.

Table 11.1 Parameter estimates and their standard error

	β_1	β_2	β_{Rolling}	$\beta_{\text{Mountainous}}$
1. $\hat{E}\{\mu\} = \beta_0 X_1{}^{\beta_1} X_2{}^{\beta_2}$	1.115 ± 0.061	1.033 ± 0.030		
2. $\hat{E}\{\mu\} = \beta_0 X_1{}^{\beta_1} X_2{}^{\beta_2} \beta_{\text{Terrain}}$	1.028 ± 0.027	0.947 ± 0.023	1.595 ± 0.018	2.375 ± 0.026

[11] See Sect. 6.6.

[12] The $\pm$one standard error notation is common in the experimental sciences and in engineering.

[13] See Sect. 9.6.

[14] To download this spreadsheet go to http://extras.springer.com/ and enter the ISBN of this book. Look in the "Spreadsheets" folder for "Chap. 11. Simulation $X1$, $X2$ and Terrain.xls or xlsm."

[15] See Sect. 9.1.2.

11.3.3 How Accuracy Is Affected by the AADT "Error in Variables"

The X_2 variable in the model equations is the normalized AADT. The AADT data are based on short-term traffic counts conducted during few days of the year once every few years. These counts are then "factored-up" to a year taking into account the specific hours of day and days of year of the short-term traffic count. The record at some permanent counting station where traffic is counted throughout the year and which is thought representative of that road segment is used for this purpose. These factored-up AADT estimates are then interpolated for years in which there was no short-term traffic count.

Naturally the AADT data are subject to considerable uncertainty. How large the uncertainty is depends on the details of jurisdiction-specific practices. Simulation was used to examine the effect of this "error-in-variable" on the accuracy with which $E\{\mu\}$ is estimated.[16] The assumption was that the AADTs are drawn from a Gamma distribution the mean of which is that of the Colorado data[17], and the standard deviation of which is once 10 % and another time 30 % of that mean. The results are summarized in Fig. 11.4.

Thus, the uncertainty surrounding the AADT estimates has a substantive effect on the accuracy with which the $E\{\mu\}$ is estimated.

What the simulation says about how the accuracy of parameter estimates depends on the inaccuracy of AADT is shown in Table 11.2. Surprisingly, the inaccuracy of AADT does not seem to have a large adverse effect on the standard error of parameter estimates.

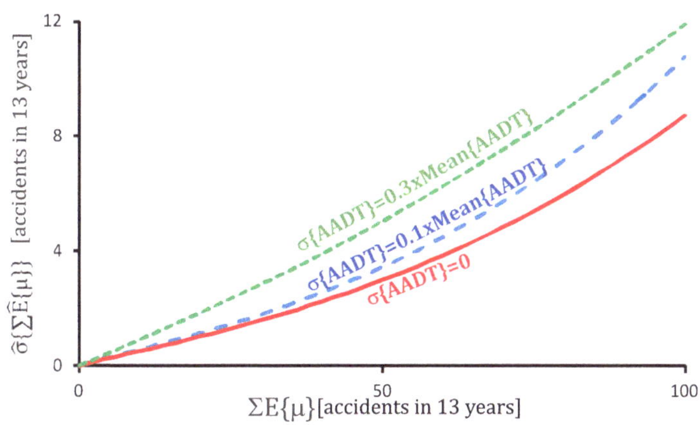

Fig. 11.4 The effect of AADT "error in variables"

[16] To download this spreadsheet go to http://extras.springer.com/ and enter the ISBN of this book. Look in the "Spreadsheets" folder for "Chap. 11. Simulation, Error in AADT.xls or xlsm."

[17] See Fig. 3.1.

Table 11.2 Parameter estimates and their standard error for $\hat{E}\{\mu\} = \beta_0 X_1^{\beta_1} X_2^{\beta_2} \beta_{\text{Terrain}}$

	β_1	β_2	β_{Rolling}	$\beta_{\text{Mountainous}}$
1. $\sigma\{\text{AADT}\} = 0$	1.028 ± 0.027	0.947 ± 0.023	1.595 ± 0.018	2.375 ± 0.026
2. $\sigma\{\text{AADT}\} = 0.1 \times \text{Mean}\{\text{AADT}\}$	1.064 ± 0.028	0.950 ± 0.023	1.611 ± 0.018	2.346 ± 0.026
3. $\sigma\{\text{AADT}\} = 0.3 \times \text{Mean}\{\text{AADT}\}$	1.071 ± 0.098	0.949 ± 0.106	1.604 ± 0.020	2.364 ± 0.080

11.3.4 Study Design

The question often arises how much data is needed to develop a satisfactory SPF. When the focus is on practical applications[18] the response is in terms of the accuracy of $\hat{E}\{\mu\}$ and the magnitude of $\sigma\{\mu\}$. When the focus is on research one is interested in the accuracy of the parameter estimates.

The data used throughout the book pertains to 5,323 Colorado two-lane rural road segments totaling 6,029 miles, on which, during the 13 years between 1986 and 1998, 21,718 Injury & Fatal occurred. Is this an overkill? One can ask, e.g., how large would be the standard error of the estimates of $E\{\mu\}$ or of the β's if, say, only 3 years of data were used. These are the kinds of questions that arise when the development of an SPF is first planned. Simulation can be used to answer. Here the task is to determine the standard error of $\hat{E}\{\mu\}$ if the most recent 1, 3, or 5 years of data were used instead of all 13. The model equation was $\hat{E}\{\mu\} = \beta_0 X_1^{\beta_1} X_2^{\beta_2} \beta_{\text{Terrain}}$ and $\sigma\{\text{AADT}\} = 0.1 \times \text{Mean}\{\text{AADT}\}$ was used. The results are in Fig. 11.5.[19]

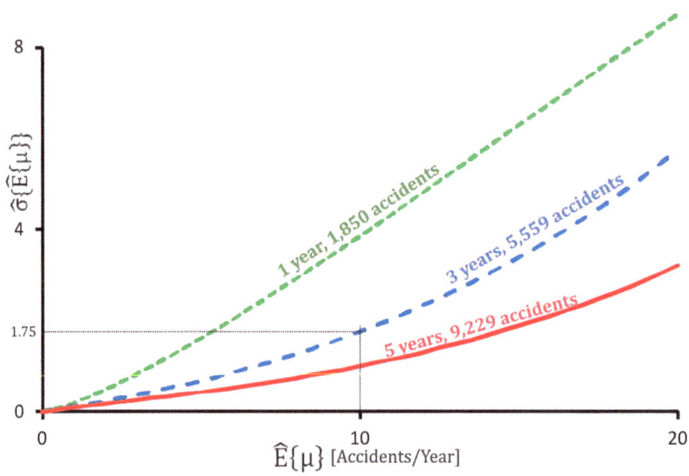

Fig. 11.5 The standard error of $\hat{E}\{\mu\}$ vs. data years or number of accidents

[18] See Sect. 1.5.

[19] To download this spreadsheet go to http://extras.springer.com/ and enter the ISBN of this book. Look in the "Spreadsheets" folder for "Chap. 11. Simulation, Years of data.xls or xlsm"

Thus, for example, with 5,559 accidents in the most recent 3 years of Colorado data, for a population of sites where the average is 10 accidents/year, the standard error of the estimate of $E\{\mu\}$ which the model produces is about ± 1.75. The standard proviso is that this would be a valid estimate if $\hat{E}\{\mu\} = \beta_0 X_1^{\beta_1} X_2^{\beta_2} \beta_{\text{Terrain}}$ was the right model equation, if no safety-related variables were missing from it, and if minimizing $\sum|\text{residuals}|$ was the correct objective for parameter estimation.

How the accuracy of parameter estimates depends on the number of accidents in the data is shown in Table 11.3.

Table 11.3 Parameter estimates and their standard errors

	β_1	β_2	β_{Rolling}	$\beta_{\text{Mountainous}}$
1998, 1 year, 1,850 accidents	1.435 ± 0.063	1.261 ± 0.049	4.057 ± 2.669	7.952 ± 5.336
1996–1998, 3 years, 5,559 accidents	1.163 ± 0.044	1.104 ± 0.073	1.865 ± 0.247	3.112 ± 0.500
1994–1998, 5 years, 9,229 accidents	1.171 ± 0.047	1.083 ± 0.053	1.418 ± 0.141	2.253 ± 0.211
1986–1998, 13 years, 21,718 accidents	1.064 ± 0.027	0.950 ± 0.023	1.611 ± 0.018	2.346 ± 0.026

11.4 Summary

The "goodness" of an estimate is not measured by its fit to data; it is measured by how close it is to the real value. Since the aim of SPF modeling is to produce good estimates, estimate accuracy is the touchstone for modeling choices and decisions.

The problem is that in regressions of this kind the determination of estimate accuracy is problematic. The difficulty is conceptual. Accuracy can be quantified only in a conditional manner; that is, we can say what would be if some conditions held. The conditions are that: (a) the model equation used is the right one, (b) the objective function that is optimized returns the right parameter estimates, and (c) no important safety-related predictor variables are missing from the model equation. Not only are these conditions not met, these are some of the key modeling decisions which the determination of estimate accuracy should help us to make. It follows that such a conditional estimate of accuracy is overly optimistic and that its very conditionality begs the question: How can one use estimation accuracy to decide between competing model equations or objective functions if that accuracy depends on the model equation or objective function assumed to be correct?

In this chapter this conditional accuracy was determined by simulation. The main idea was to assume that the results of modeling are the true values and then to use realistic and explicit mechanisms for introducing randomness into data. Thus, for example, μ's were assumed to come from a Gamma distribution, accident counts from a Poisson, AADT estimates were assumed to have a stated variance, etc. Data generated in this manner were then used in the C-F spreadsheet to give rise to new estimates of $E\{\mu\}$, $\sigma\{\mu\}$, and model parameters. From the results of few such simulation runs, the standard error of estimates could be ascertained.

The VBA simulation code has been used to examine several issues: how accuracy is affected by the addition of a variable, how the variance of AADT estimates influences the accuracy of $\hat{E}\{\mu\}$, and what is the effect on that accuracy of using fewer years of data.

References

Wood GR (2005) Confidence and prediction intervals for generalized linear accident models. Accid Anal Prev 37:267–273

Closure 12

I set out in pursuit of two objectives: to show how to develop Safety Performance Functions using only a common spreadsheet and to lay the foundations for an understanding of statistical regression modeling in road safety. I hope that progress has been made towards both goals.

When the suggestion that a lowly spreadsheet can be used to advantage was first made it may have seemed outlandish. But by now it is evident that the tool fits the aims. The art of modeling is best presented with all its elements in plain sight and the spreadsheet is an admirable environment in this way. The modeler has to make choices that require insight and understanding and the construction of the C-F spreadsheet relies on such an understanding and promotes it. After all, it is "Cogito," not "Computo ergo sum."[1]

The most common aspiration of modelers is to develop SPFs that allow one to say by how much the $E\{\mu\}$ might change if by some manipulation one changed the value of a predictor variable. I (and many others) doubt that regression models based on observational cross-section data can deliver such goods.[2] This is why I settled for the modest and largely uncontroversial aim of developing SPFs that provide good estimates of the $E\{\mu\}$ and $\sigma\{\mu\}$. This election shaped the modeling approach to a surprising extent. The main consequence was a shift of attention from the plumage to the bird; from preoccupation with the parameters to the quality of the dependent variable.

For didactic purposes, the chapter topics were a succession of building blocks, each dealing with a specific subject matter: how to do an EDA, how to fit a function to data, how to examine the quality of a fit, etc. These subjects occasionally required the introduction of a tool: the Pivot Table for EDA, the SOLVER for parametric curve-fitting, the CURE plot for fit quality. It is my hope that, taken together, the

[1] The Latin phrase *cogito ergo sum* (I think, therefore I am) is a philosophical proposition by René Descartes.

[2] For reasons see Sects. 6.7 and 10.2.

E. Hauer, *The Art of Regression Modeling in Road Safety*,
DOI 10.1007/978-3-319-12529-9_12

building blocks and the tools amount to a coherent guide to the art and practice of SPF development.

As construed here, modeling is not of a once-through process in which, once the data are assembled, the rest is up to some comprehensive and automated software. The picture painted is of a modeler who has to make choices: what to optimize, what variables to use, by what functions to represent their influence, how to purge bias from fits, etc. If this modeling process was pictured as made up of sequentially ordered activities, it might be this: a variable is considered for introduction into the model equation → the function by which its influence is accounted for is chosen → the parameters of the function are estimated → the resulting CURE plots are examined → alternative functions that could improve the CURE plots are tested. These activities make for one turn of the modeling spiral in Fig. 12.1. The next cycle begins by considering the introduction of another variable. Occasionally, as in the "snakes-and-ladders" game one has to go back a turn or two to try another option or correct a misstep. The spiral has no natural end. The process reaches its end when all available data has been used and when no opportunity for improvement is evident.

Modeling is seen as a sequence of choices; choice involves consideration; and consideration requires on understanding. This book with all its warts is a witness to my incomplete understanding. I hope that the right issues were raised even when the right answers could not always be given. I also hope that this unconventional view of regression modeling will prompt others to search for better ways.

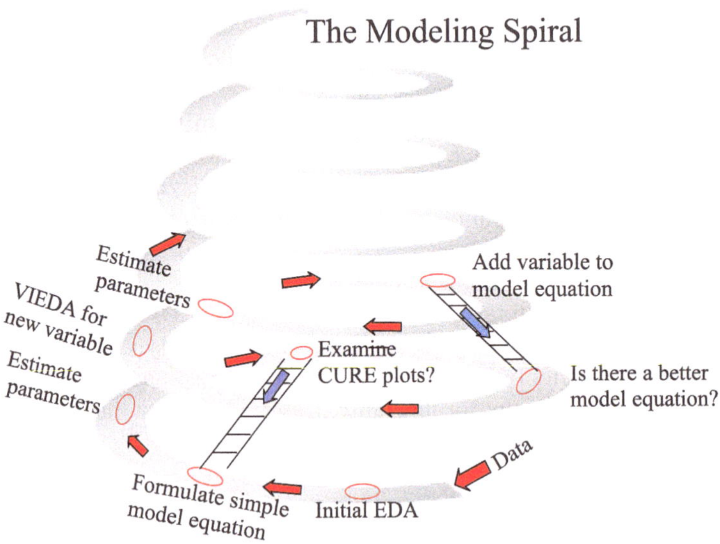

Fig. 12.1 A schematic representation of the modeling process

Appendices

The appendices contain various materials mentioned in the main text and about which the interested reader may want to have more detail.

Appendix A: Accident Counts on a Unit: The Poisson Assumption

The common assumption is that accident counts for an unchanging unit are approximately Poisson distributed. That this is a serviceable approximation has been assumed for a long time.[1] The use of the Poisson distribution for accident counts may have started with L. von Bortkiewitcz[2] who had data about the number of deaths by horse-kick in ten Prussian army corps over 20 years. When comparing the number of years with 0, 1, . . ., 5 deaths to the number predicted by the Poisson distribution the fit was remarkably good.[3]

There have been attempts to examine whether road accident counts actually obey the Poisson distribution.[4] A time series of yearly accident counts for 35 urban intersections served as data. The original analysis led to the conclusion that there are grounds for doubting the general validity of the Poisson assumption. However, when the data were later re-analyzed using a better statistical test the conclusion was that "there is little (if any) justification for rejecting the Poisson distribution."

The time-honored assumption that accident counts obey the Poisson distribution is impossible to refute by data. The reason is that to examine whether a set of accident counts fits the Poisson distribution, one would need many counts which all materialized while the same mean number of accidents per unit of time (λ) prevailed. However, on real transportation systems the conditions that determine the λ change over time. Therefore, support for the conjecture that accident counts

[1] See, e.g., Gerlough (1955).

[2] Von Bortkiewitcz (1898).

[3] There were later claims, perhaps unfounded, that the data has been censored to improve the fit. But the essential reason for the good fit was that the numbers were small and the variation in the mean (λ) from year-to-year and from one army corps to another was small (Quine and Seneta 1987).

[4] Nicholson (1985, 1986) and Nicholson and Wong (1993).

E. Hauer, *The Art of Regression Modeling in Road Safety*,
DOI 10.1007/978-3-319-12529-9

are Poisson distributed must come mainly from the conviction that if λ did remain constant then the nature of the accident generation process would approximate the conditions which, by logical necessity, give rise to the Poisson distribution. That this is a reasonable conviction to hold is argued below.

Imagine a period of time during which all factors affecting the safety of some unit (traffic flow, weather, geometry, traffic control, road user demography, etc.) as well as the factors affecting the reporting of accidents remained nearly constant. If so the mean number of reported accidents per unit of time, λ would be constant.[5] Imagine further that this time period is finely divided (as in Fig. A.1) so that within each subdivision of duration Δt the probability of one accident is $\lambda \Delta t$ and the probability of more than one accident within Δt is negligibly small.[6]

To illustrate, consider an intersection where the mean accident frequency λ is 0.006 accidents/h. If $\Delta t = 1$ min, the probability of an accident in Δt is $0.006/60 = 10^{-4}$. Were the occurrence and reporting of one accident entirely independent of the occurrence or nonoccurrence and reporting of another, the chance of two accidents occurring in the same minute would be 1×10^{-8} or once in about 200 years. If our interest is in the distribution of the monthly or yearly accident counts, the neglect of one accident in 200 years is immaterial for all practical purposes. Even if the assumption of independence was not true, as is likely, and if the chance of a second accident to occur once the first has occurred was, say, ten times more than that of the first one (i.e., 10^{-3}), one might find two accidents in the same minute only once in 20 years.

Under the aforementioned conditions the occurrence of accidents is said to be a "Poisson Process." A purely logical argument will show[7] that for a Poisson Process the probability (P) that the number of accidents (K) in time period T during which λ prevails is k, is given by the Poisson distribution:

$$P(K = k|\mu) = \frac{e^{-\mu}\mu^k}{k!} \quad \text{where } \mu = \lambda T \tag{A.1}$$

In this equation μ is the number of accidents expected to be reported during T. Thus, for example, in $T = 100$ h during which $\lambda = 0.006$ accidents/h (and therefore $\mu = 0.6$

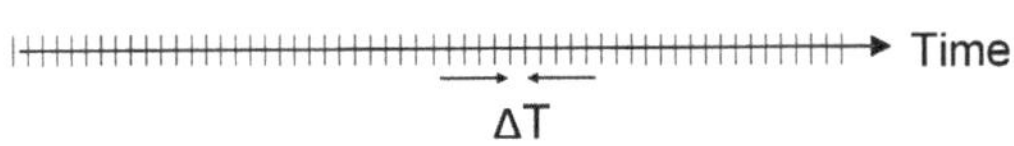

Fig. A.1 Fine subdivision of the time axis

[5] The dimension of λ is the reciprocal of time.

[6] More formally, let $K_{t,\, t+\Delta t}$ be the number of accidents occurring in the time interval $(t, t+\Delta t]$ where $\Delta t > 0$. The conditions which lead to a Poisson process are (1) $P(K_{t,\, t+\Delta t} = 0) = 1 - \lambda\Delta t + o(\Delta t)$, (2) $P(K_{t,\, t+\Delta t} = 1) = \lambda\Delta t + o(\Delta t)$, (3) $P(K_{t,\, t+\Delta t} \geq 2) = o(\Delta t)$, and (4) that $K_{t,\, t+\Delta t}$ is statistically independent of the number and time of accidents that occurred in $(0,t]$. The expression $o(\Delta t)$ denotes a function any $f(\Delta t)$ which goes to 0 more rapidly than Δt (i.e., $\lim_{\Delta t \to 0} \frac{f(\Delta t)}{\Delta t} = 0$).

[7] See, e.g., Sect. 5.7 in Freund and Walpole (1987).

accidents in 100 h), $P(K=0)=e^{-0.6}=0.549$, $P(K=1)=P(K=0)\times 0.6/1=0.329$, $P(K=2)=P(K=1)\times 0.6/2=0.099, \ldots$.[8]

The Poisson distribution has two convenient properties:

Property 1: If K is Poisson distributed then $E\{K\}=V\{K\}$.

Property 2: If the accident counts $K_1, K_2, \ldots, K_n$ are statistically independent and Poisson distributed with means $\mu_1, \mu_2, \ldots, \mu_n$ then their sum is also Poisson distributed with a mean of $\mu_1+\mu_2+\cdots\mu_n$.

Appendix B: The Poisson Likelihood Function

For units 1, 2, …, i, …, n the accident counts are $k_1, k_2, \ldots, k_i, \ldots, k_n$. If the accident counts are independent and Poisson distributed with means $\mu_1, \mu_2, \ldots, \mu_i, \ldots, \mu_n$ then by the "multiplication rule for independent events" the likelihood function is:

$$\mathcal{L}(\mu_1,\ldots,\mu_n)\prod_{i=1}^{n}\frac{e^{-\mu_i}{\mu_i}^{k_i}}{k_i!} \tag{A.2}$$

Because logarithms preserve the location of the maximum and since the $\prod_1^n k_i!$ is a constant that does not depend on any parameter it is sufficient to seek the maximum of $\ln(\mathcal{L}^*)$, the "abridged" log-likelihood:

$$\ln[\mathcal{L}^*(\mu_1,\ldots,\mu_n)]=\sum_{i=1}^{n}(-\mu_i)+k_i\ln(\mu_i) \tag{A.3}$$

In a model equation the μ's are functions of variable values and parameters. Replacing the μ's in (A.3) by the model equation makes the abridged log-likelihood into a function of parameters. Thus, for example, we examined the model equation $\mu_i=\beta_0(\text{Segment Length}_i)^{\beta_1}$ in which "Segment Length" is the variable and β_0, β_1 are the parameters. This makes every μ_i into a function of β_0 and β_1 and thereby $\ln[\mathcal{L}^*\mu_1,\ldots,\mu_n]$ turns into $\ln[\mathcal{L}^*(\beta_0,\beta_1)]$. Fitting this model equation to data amounts to finding those values of β_0 and β_1 which maximize the $\ln[\mathcal{L}^*(\beta_0,\beta_1)]$.

Appendix C: The Variance of μ's and of Accident Counts in a Population of Units

The foundational axiom was that the μ of a unit is determined by its safety-related traits. Since the units in a population share some such traits but differ in many others, the corollary of the foundational axiom is that their μ's must differ.

[8] The recursive relationship $P(K=k)=P(K=k-1)\times\mu/k$ is convenient in computations.

This diversity of μ's is illustrated in the upper tier of Fig. A.2[9] by the different μ's of five units comprising a population. The lower tier of that figure shows the count of accidents on each unit.

The diversity of the unobserved μ's in the upper tier can be measured by $V\{\mu\}$ and that of the observed k's by $V\{K\}$. The question is how $V\{\mu\}$ and $V\{K\}$ are related and how the estimate of $V\{K\}$ can be used to estimate $V\{\mu\}$.

To establish the relationship between $V\{\mu\}$ and $V\{K\}$ a convenient starting point is the law of total variance. This logical relationship states that for random variables X and Y

$$V\{Y\} = E\{V\{Y|X\}\} + V\{E\{Y|X\}\} \tag{A.4}$$

If μ takes the place of X and K of Y, the inner part of the first term on the right hand side, the $V\{Y|X\}$, is the variance of the accident count K_i for a unit with the mean μ_i. Assuming that accident counts are Poisson distributed $V\{K_i|\mu_i\} = \mu_i$ and therefore the first summand is $E\{\mu\}$. The inner expression of the second term of the right hand side, the $Y\{Y|X\}$, is the mean value of K_i when the mean is μ_i. Assuming again that the accident counts are Poisson distributed, $E\{K_i|\mu_i\} = \mu_i$ and therefore the second summand is $V\{\mu\}$. It follows that

$$V\{K\} = E\{\mu\} + V\{\mu\} \quad \text{or} \quad V\{\mu\} = V\{K\} - E\{\mu\} \tag{A.5}$$

Replacing $V\{K\}$ by the sample variance of accidents counts in the population and $E\{\mu\}$ by the sample mean of accident counts yields a "method of moments" estimate of $V\{\mu\}$.

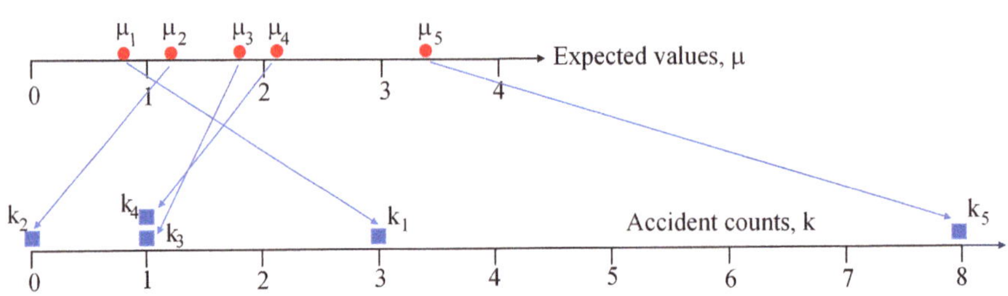

Fig. A.2 The μ's and k's in a population of units

[9] The same figure was used earlier as 1.3 and 2.2.

Appendix D: The Negative Binomial Distribution and the Gamma Assumption

The NB distribution for accident counts in a population of units is based on three assumptions:

(i) That the accident counts on every unit are Poisson distributed.
(ii) That the μ's of the units of which a population is comprised are different.
(iii) That the frequency of μ's in the population can be well approximated by a Gamma distribution.

The plausibility of assumption (i) comes from the similarity between the realities of accident occurrence and the conditions from which the Poisson distribution is derived by logic as shown in Appendix A. Assumption (ii) is plausible because units (roads, intersections, drivers) that have the same safety-related traits which define a population still differ in many other safety-related traits. Assumption (iii) has no real source of plausibility; there is no reason why the μ's in populations of units should form a Gamma distribution. The motivations for its use are convenience and flexibility. Convenience – because together with (i) and (ii) it implies a closed-form probability distribution: the Negative Binomial[10]; flexibility – because the Gamma pdf can take on many shapes as is shown in Fig. A.3.

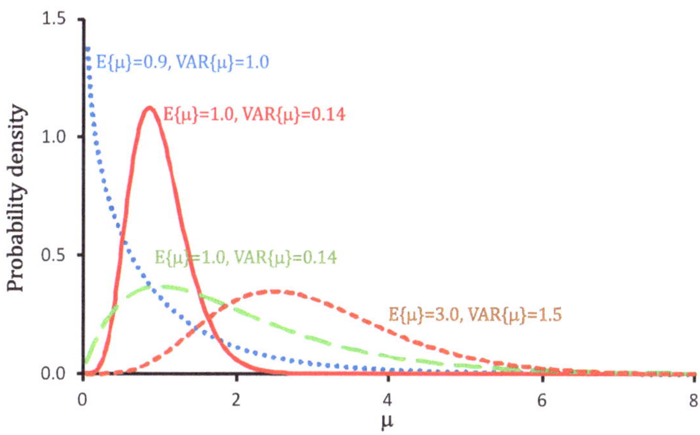

Fig. A.3 The many shapes of the Gamma probability density function

[10] The Negative Binomial distribution can arise in number of ways; one is as a continuous mixture of a Poisson distribution $\left(P(K=k)=\frac{e^{-\mu}\mu^k}{k!}\right)$ the μ in which is Gamma distributed ($f(\mu)= a^b\mu^{b-1}e^{-a\mu}/\Gamma(b)$). In this case, $P(K=k)=\int_0^\infty \frac{e^{-\mu_\mu k}}{k!}\frac{a^b\mu^{b-1}e^{-a\mu}}{\Gamma(b)}d\mu=\frac{a^b}{\Gamma(b)k!}\int_0^\infty e^{-\mu(a+1)}\mu^{k+b-1}d\mu=\frac{\Gamma(k+b)}{\Gamma(b)k!}\frac{a^b}{(a+1)^{k-b}}$. For this distribution $E\{K\}=b/a$ and $V\{K\}=b/a+b/a^2$.

Even so, there are limitations. Thus, for example, if the distribution of μ's in some population of units has more than one peak that cannot be well represented by the Gamma pdf.

The Gamma probability density function has two parameters, "a" and "b."[11]

$$f(\mu) = \frac{a^b \mu^{b-1} e^{-a\mu}}{\Gamma(b)} \tag{A.6}$$

In terms of these parameters the mean and the variance of this distribution are:

$$E\{\mu\} = \frac{b}{a} \quad \text{and} \quad V\{\mu\} = \frac{b}{a^2} = \frac{E\{\mu\}^2}{b} \tag{A.7}$$

From here,

$$a = \frac{E\{\mu\}}{V\{\mu\}} \quad \text{and} \quad b = aE\{\mu\} \tag{A.8}$$

For a given $E\{\mu\}$, the $V\{\mu\}$ in (A.7) is proportional to $1/b$. That is, the larger is "b" the lesser is $V\{\mu\}$. This is why the "$1/b$" is called the "overdispersion" parameter.

Appendix E: The Negative Binomial Likelihood Function

On the foundation of assumptions (i)–(iii) in Appendix D one can build the NB likelihood function as follows. We have data about units 1, 2, . . ., i, . . ., n. Begin by considering unit "i." The traits of unit "i" that are in the model equation define population "i" of units with the same traits. We do not know the μ of the unit for which we have data but, by assumption (iii), it is one from a Gamma distribution with parameters a_i and b_i. If so, the distribution of accident counts in population "i" is NB[12]:

$$P(K_i = k_i) = \frac{\Gamma(k_i + b_i)}{\Gamma(b_i)k_i!} \frac{a_i^{b_i}}{(a_i + 1)^{k_i+b_i}} \tag{A.9}$$

[11] The "a" is the so-called "rate" (or reciprocal of "scale") parameter and "b" is the "shape" parameter. The Excel spreadsheet function is GAMMADIST(μ, b, $1/a$, TRUE for cumulative or FALSE for density)

[12] See derivation in Appendix D and earlier mention in (1.6) and (8.1).

With this, the likelihood function is:

$$\mathcal{L}[(a_1,b_1),\ldots,(a_i,b_i),\ldots,(a_n,b_n)] = \prod_{i=1}^{n} \frac{\Gamma(k_i+b_i)}{\Gamma(b_i)k_i!} \frac{{a_i}^{b_i}}{(a_i+1)^{k_i+b_i}} \tag{A.10}$$

At this point two considerations are invoked. First, because $E\{\mu_i\} = b_i/a_i$, the a_i in (A.10) can be replaced by $b_i/E\{\mu_i\}$. With this substitution

$$\mathcal{L}(b_1,\ldots,b_n,E\{\mu_i\}) = \prod_{i=1}^{n} \frac{\Gamma(k_i+b_i)}{\Gamma(b_i)k_i!} \left(\frac{b_i}{b_i+E\{\mu_i\}}\right)^{b_i} \left(\frac{E\{\mu_i\}}{b_i+E\{\mu_i\}}\right)^{k_i} \tag{A.11}$$

The second consideration pertains to the "shape" parameter of the Gamma pdf. It can be shown[13] that for road segments that differ in length to have the correct influence in parameter estimation, and also for coherence when SPFs are used to estimate the μ of a road segment, it is best to use

$$b_i = \mathscr{b} \times L_i \text{ where } \mathscr{b} \text{ is a parameter and } L_i \text{ is the length of segment } "i" \tag{A.12}$$

Equation (A.12) amounts to adding assumption (iv)[14], namely that the shape parameter is always proportional to the Segment Length and that there is a constant of proportionality ($\mathscr{b}$) that is common to all the populations. While assumptions (i) and (ii) are justifiable on logical grounds and assumption (iii) has the merit of being flexible and leading to a closed-form expression for the distribution of accident counts, it is difficult to offer a convincing justification for assumption (iv).

Replacing $E\{\mu_i\}$ with $\hat{E}\{\mu_i\}$, $\mathcal{L}$ becomes a function of the parameters $\beta_0, \beta_1, \ldots$ and of $\mathscr{b}$. Now the abridged log-likelihood is

$$\begin{aligned}\ln[\mathcal{L}^*(\beta_0,\beta_1,\ldots,\mathscr{b})] = \sum_{i=1}^{n} [&\ln\Gamma(k_i+\mathscr{b}L_i) - \ln\Gamma(\mathscr{b}L_i) + \mathscr{b}L_i\ln(bL_i) \\ &+ k_i\ln\left(\widehat{E}\{\mu_i\}\right) - (\mathscr{b}L_i+k_i)\ln\left(\mathscr{b}L_i+\widehat{E}\{\mu_i\}\right)]\end{aligned} \tag{A.13}$$

This is the abridged NB log-likelihood function used in Chap. 8.

Although the Poisson likelihood function is simpler, its drawback is that it is inconsistent with the logical necessity and empirical reality of overdispersion. Being able to account for overdispersion is the main merit of the NB likelihood function. It also has the practical advantage of simultaneously providing estimates of $E\{\mu\}$ and $V\{\mu\}$. These desirable features come at a price: the need to make specific and difficult-to-justify assumptions about the distribution of μ's in populations and about $\mathscr{b}$ being a constant for all populations. Even though the mathematical deductions are solid, the end result is only as good as the assumptions on which it rests.

---

[13] See Hauer (2001).

[14] If the units are intersections $L = 1$.

Appendix F: The Conditional Expectation, $E\{\mu|K=k\}$

The SPF provides an estimate of $E\{\mu\}$, the mean μ's in a population of units with specified traits. Consider a subpopulation of those units which all recorded k accidents. What is the estimate of the mean of μ's (denoted by $E\{\mu|K=k\}$) of this subpopulation? This is needed to answer questions such as those posed in Sects. 1.4.4 and 1.4.5.

By the Bayes Theorem for distributions,

$$f(\mu|K=k) = (\text{constant})P(K=k|\mu)f(\mu) \tag{A.14}$$

Substituting the corresponding expressions from (A.1) and (A.6),

$$\begin{aligned} f(\mu|K=k) &= (\text{constant } 1)\frac{\mathrm{e}^{-\mu}\mu^k}{k!}\frac{a^b\mu^{b-1}\mathrm{e}^{-a\mu}}{\Gamma(b)} \\ &= (\text{constant } 2)\mu^{k+b-1}\mathrm{e}^{-\mu(1+a)} \end{aligned} \tag{A.15}$$

After integrating over all $\mu > 0$, the constant needed to make the integral into 1 is $\frac{(1+a)^{k+b}}{\Gamma(k+b)}$. It follows that:

$$f(\mu|K=k) = \frac{(1+a)^{k+b}\mu^{k+b-1}\mathrm{e}^{-\mu(1+a)}}{\Gamma(k+b)} \tag{A.16}$$

This too is a Gamma pdf except that "a" turned into $a+1$ and "b" into $k+b$. In correspondence, the mean and variance of the conditional distribution are:

$$E\{\mu|K=k\} = \frac{k+b}{a+1} \quad \text{and} \quad V\{\mu|K=k\} = \frac{k+b}{(a+1)^2} \tag{A.17}$$

These are the expressions used in Sect. 1.4.4.

Appendix G: The Negative Multinomial Likelihood Function

The NB likelihood function can be used when the data pertain to a single period of time. Until late in Chap. 8 the 5 years 1994–1998 were considered to be a single period; the accident count was the sum for those 5 years and Annual Average Daily Traffic (AADT) was the period average. The original Colorado data give accident counts and AADT estimates for each of 13 years. When data is for multiple time periods[15] the NB likelihood function is replaced by the Negative Multinomial (NM); its derivation is given below.[16]

[15] If in a cross-sectional sample for the same units the observed traits are available for two or more periods the data are said to form a panel, longitudinal, or cluster data set.

[16] The derivation follows Guo (1996).

For unit i we have data for time periods 1, 2, . . ., j, . . ., m_i. The mean accident count for unit "i" in time period "j" is $\mu_{i,j}$. The traits of unit "i" in period "j" define an imagined population of units. The μ's of this population are assumed to be Gamma distributed with a mean $E\{\mu_{i,j}\}$ and variance $E\{\mu_{i,j}\}^2/b$. Or, equivalently,

$$\mu_{i,\,j} = E\{\mu_{i,\,j}\}\theta_i \tag{A.18}$$

where θ_i comes from the Gamma distribution the density of which is

$$f(\theta_i) = \frac{\theta_i^{b-1}\mathrm{e}^{-b\theta_i}b^b}{\Gamma(b)} \tag{A.19}$$

The mean of this pdf is 1 and its variance is $1/b$.

Equation (A.18) contains a strong assumption. Namely that the μ's of all units in population i change from one period to another in proportion to the change in mean of the μ's in that population.

If the $k_{i,1}$, $k_{i,2}$, . . ., $k_{i,j}$, . . ., $k_{i,mi}$ accident counts for unit i over the m_i periods are independent and Poisson distributed and when θ_i is given, the probability to observe these counts is:

$$P\left(k_{i,1}, k_{i,2}, \ldots, k_{i,j}, \ldots, k_{i,m_i} \middle| \theta_i\right) = \prod_{j=1}^{m_i} \frac{\left(E\{\mu_{i,j}\}\theta_i\right)^{k_{i,j}}\mathrm{e}^{-E\{\mu_{i,j}\}\theta_i}}{k_{i,j}!} \tag{A.20}$$

However, if θ_i is not given but known is the pdf in (A.19) then the probability to observe the $k_{i,1}$, $k_{i,2}$, . . ., $k_{i,j}$, . . ., $k_{i,mi}$ accident counts is:

$$P\left(k_{i,1}, k_{i,2}, \ldots, k_{i,j}, \ldots, k_{i,m_i}\right) = \int_0^\infty \prod_{j=1}^{m_i} \frac{\left(E\{\mu_{i,j}\}\theta_i\right)^{k_{i,j}}\mathrm{e}^{-E\{\mu_{i,j}\}\theta_i}}{k_{i,j}!} \frac{\theta_i^{b-1}\mathrm{e}^{-b\theta_i}b^b}{\Gamma(b)}\, d\theta_i \tag{A.21}$$

For notational brevity, I will use

$$k_i \equiv \sum_{j=1}^{m_i} k_{i,j}, \quad E\{\mu_i\} \equiv \sum_{j=1}^{m_i} E\{\mu_{i,j}\}, \quad \text{and} \quad \prod_{j=1}^{m_i} \theta_i^{\,k_{i,j}} = \theta_i^{\sum_{j=1}^{m_i} k_{i,}}$$
$$\equiv \theta_i^{\,k_i} \tag{A.22}$$

With this notation (A.21) can be rewritten as

$$P\left(k_{i,1}, k_{i,2}, \ldots, k_{i,j}, \ldots, k_{i,m_i}\right) = \frac{b^b\left(\prod_{j=1}^{m_i} E\{\mu_{i,\,j}\}^{k_{i,j}}\right)}{\left(\prod_{j=1}^{m_i} k_{i,\,j}!\right)\Gamma(b)}$$
$$\int_0^\infty \theta_i^{k_i+b-1}\mathrm{e}^{-\theta_i(E\{\mu_i\}+b)}d\theta_i = \frac{b^b\left(\prod_{j=1}^{m_i} E\{\mu_{i,\,j}\}^{k_{i,j}}\right)\Gamma(k_i+b)}{\left(\prod_{j=1}^{m_i} k_{i,\,j}!\right)\Gamma(b)(E\{\mu_i\}+b)^{k_i+b}} \tag{A.23}$$

As noted in Appendix E to preserve consistency in EB estimation b is replaced by $b_i = \ell L_i$. Replacing $E\{\mu_{i,j}\}$ with $\hat{E}\left\{\mu_{i,j}\right\}$ of the model equation the left hand side of the equation turns it into a function of the parameters $\beta_0, \beta_1, \ldots$ and ℓ. Viewed in this way it is the contribution of unit i to the likelihood function. Omitting the constant with the $k_{i,j}!$, the contribution of unit i to the abridged log-likelihood is

$$\begin{aligned}\ln\left[\mathcal{L}_i^*(\beta_0, \beta_1, \ldots, \ell)\right] = {} & \ell L_i \ln(\ell L_i) + \left[\sum_{j=1}^{m_i} k_{i,j} \ln\left(\widehat{E}\{\mu_{i,j}\}\right)\right] \\ & + \ln\Gamma\left[\left(\sum_{j=1}^{m_i} k_{i,j}\right) + \ell L_i\right] - \ln\Gamma(\ell L_i) \\ & - \left[\left(\sum_{j=1}^{m_i} k_{i,j}\right) + \ell L_i\right] \ln\left[\left(\sum_{j=1}^{m_i} \widehat{E}\{\mu_{i,j}\}\right) + \ell L_i\right]\end{aligned} \tag{A.24}$$

The maximum likelihood parameters are those that maximize

$$ln(\mathcal{L}^*) = \sum_{i=1}^{n} ln\left[\mathcal{L}_i^*(\beta_0, \beta_1, \ldots, \ell)\right] \tag{A.25}$$

Appendix H: The Nadaraya Watson Kernel Regression

In two dimensional smoothing we have a set of "points" given by coordinate pairs (X_i, Y_i), $i = 1, 2, \ldots, N$. Variables values are on the abscissa (X). The N-W estimator at $X = x$ is a weighted average of the Y's:

$$\hat{y}(x) = \sum_{1}^{N} w_i Y_i \tag{A.26}$$

The weight, w_i is given by

$$w_i = \frac{K\left(\frac{x - X_i}{h}\right)}{\sum_1^N K\left(\frac{x - X_i}{h}\right)} \tag{A.27}$$

$K(.)$ is the "kernel function" and h the "bandwidth."

There are several popular kernel functions. In this book the Gaussian Kernel was used. The Gaussian kernel is:

$$K(.) = \mathrm{e}^{-\frac{(x - X_i)^2}{2h^2}} \tag{A.28}$$

in which the bandwidth (h) corresponds to the standard deviation of the Gaussian (Normal) probability distribution.

So that the Gaussian window not go beyond the data, "h" is variable near the edges; it is chosen to make sure that $(x - 0)/h$ and $(x_N - x)/h$ are larger than, say, 3. To illustrate, suppose that x measures Segment Length which varies from 0 to 10 miles and that away from the edges the bandwidth is to be 0.6 miles. If so, near the origin h must increase linearly from 0 at the origin to 0.6 at $x = 3 \times 0.6 = 1.8$ miles and decrease from 0.6 in the same manner from $x = 8.2$ miles. Thus, denoting the bandwidth away from the edges by h^*,

$$h = \begin{cases} \frac{x}{3} & \text{when} \quad x \le 3h^* \\ h^* & \text{when } 3h^* < x \le X_N - 3h^* \\ \frac{X_N - x}{3} & \text{when } x > X_N - 3h^* \end{cases} \tag{A.29}$$

For h to reach h^* it must not exceed $X_N/6$.

When smoothing is in three dimensions the data points are the coordinates (X_i, Y_i, Z_i), i = 1, 2, ..., N. The X and Y measure two variables on the ordinates, and Z is the abscissa. The only difference between two and three dimensions is that now the kernel function is

$$K(.) = \mathrm{e}^{-\frac{(x-X_i)^2}{2h_x^2} - \frac{(y-Y_i)^2}{2h_y^2}} \tag{A.30}$$

and that two bandwidths, h_x and h_y, have to be chosen.

Appendix I: The CURE Limits

Let the residuals be arranged in increasing order of an explanatory variable of interest and numbered consecutively, such that n is the total number of data points (residuals), "i" is an integer between 1 and n, and $S(i)$ is the sum of the residuals from 1 to i. If the mean of all residuals is 0, then $E\{S(i)\} = 0$. On the assumption that $E\{S(i)\} = 0$ for all i, the variance of $S(i)$, denoted by $\sigma_S{}^2(i)$, can be estimated by the sum of cumulative squared residuals at i.

Even if the individual residuals do not have the same variance, one can invoke a loose version of the central limit theorem to argue that the sum $S(i)$ is approximately normally distributed with mean = 0 and variance $\sigma_S{}^2(i)$. Similarly, the sum of the residuals for the remainder of the random walk has a mean 0 and a variance $\sigma_S{}^2(n) - \sigma_S{}^2(i)$ On this basis, the probability density function for the realization that the random walk reaches $S(i) = s$ at i and returns to 0 at n is the product of two normal probability densities. This product can be written as

$$\frac{1}{\sqrt{2\pi}\sqrt{\sigma_S^2(i)}}e^{-\frac{s^2}{2\sigma_S^2(i)}} \times \frac{1}{\sqrt{2\pi}\sqrt{\sigma_S^2(n)-\sigma_S^2(i)}}e^{-\frac{s^2}{2\left[\sigma_S^2(n)-\sigma_S^2(i)\right]}} \tag{A.31}$$

To simplify, the exponents in (A.31) can be rewritten as

$$e^{-\frac{s^2}{2\sigma_S^2(i)}}e^{-\frac{s^2}{2\left[\sigma_S^2(n)-\sigma_S^2(i)\right]}} = e^{-\frac{s^2}{2}\frac{\sigma_S^2(n)}{\sigma_S^2(i)\left[\sigma_S^2(n)-\sigma_S^2(i)\right]}} = e^{-\frac{s^2}{2\sigma_s'2(i)}} \tag{A.32}$$

where $\sigma_S'2(i) \equiv \frac{\sigma_S^2(i)\left[\sigma_S^2(n)-\sigma_S^2(i)\right]}{\sigma_S^2(n)} = \sigma_S^2(i)\left(1-\frac{\sigma_S^2(i)}{\sigma_S^2(n)}\right)$

The remaining part of (A.31) can be rewritten as

$$\frac{1}{\sigma_S(i)\sqrt{2\pi}} \times \frac{1}{\sqrt{2\pi}\sqrt{\sigma_S^2(n)-\sigma_S^2(i)^2}} = \frac{1}{\sqrt{2\pi}\sigma_S(n)}\frac{1}{\sqrt{2\pi}\sigma_S'(i)}$$

$$= \text{constant}\frac{1}{\sqrt{2\pi}\sigma_S'(i)} \tag{A.33}$$

Therefore, the probability density for a random walk that ends at 0 to pass through $S(i)=s$ is approximately normal with a mean 0 and standard deviation $\sigma_S'(i)$.

Appendix J: Towards Theory; First Steps

Parametric curve-fitting requires one to state the model equation. When the SPF is used to estimate the $E\{\mu\}$ of some population, it may be sufficient to choose the model equation by considerations of simplicity, goodness of fit, and quality of estimation. However, when the SPF is to be used to tell how changing a predictor variable on the right hand side of the model equation is likely to affect the $E\{\mu\}$ on its left hand side, this kind of guidance is insufficient. To predict the effect of cause the model equation must reflect a plausible causal theory.[17] The aim here is to take an initial step in that direction.

Consider a 1 km long road section. Let "*f*" be the expected frequency on this road section of reported accidents per lane per second, "*c*" the vehicle "concentration" – the (average) number of vehicles/(lane-km), "*p*" the probability of a vehicle to be in a crash in the next second, and "*r*" the probability of the crash to be reported. With this notation

$$f = c \times p \times r \tag{A.34}$$

Concentration (c) can be measured. The probabilities "*p*" and "*r*" depend on causal factors such as speed and headway. Speed, in turn, depends on concentration.

[17] These matters are discussed at length in Sects. 6.7 and 10.2.

Because for single-vehicle accidents "p" most likely diminishes as "c" increases, while the opposite may be true for multivehicle accidents[18] the two cases are handled separately. Subscripts "m" and "s" will be used to distinguish between the parameters that apply to the two cases.

Multivehicle Accidents

One may advance various hypotheses about the functions that may link "p" and "r" to factors such as concentration, speed, and headway. Which hypothesis is to be preferred depends on how well they are supported by data. Here it will be assumed that for multivehicle accidents "p" is a function of the average headway, $\overline{h}$; specifically, that $p \propto 1/\left(\left(\overline{h}\right)^{\alpha_\mathrm{m}}\right)$ where the parameter α_m is some positive number.[19] Since it is always true that $\overline{h} = 1/(c\overline{v})$, where $\overline{v}$ is the average speed,[20] the same assumption can also be stated as $p \propto (c\overline{v})^{\alpha_\mathrm{m}}$.

The probability "r" of an accident to be reported is assumed to depend on $\overline{v}$. The larger the speed, the more severe the crash and the more likely it is to be reported. Again many functions could be advanced. Here it will be assumed that $r \propto (\overline{v})^{\gamma_\mathrm{m}}$ where γ_m is a positive parameter and likely larger than 1. With these assumptions

$$f \propto c \times (c\overline{v})^{\alpha_\mathrm{m}} \times (\overline{v})^{\gamma_\mathrm{m}} = c^{1+\alpha_\mathrm{m}} \times (\overline{v})^{\alpha_\mathrm{m}+\gamma_\mathrm{m}} \tag{A.35}$$

Measurements indicate that "$\overline{v}$" is a function of "c" and that $\overline{v}(c)$ can be described by[21]:

$$\overline{v}(c) = \begin{cases} v_0 & \text{when } 0 < c < c_0 \\ v_0 - \beta(c - c_0) & \text{when } c_0 \le c < c_\mathrm{b} \end{cases} \tag{A.36}$$

In this v_0 is the average "free-flow" speed, c_0 is the largest concentration at which the traffic is still moving at "free-flow speed," and c_b is the concentration at which a bottleneck begins to form when the average speed is v_b. Equation (A.36) means that up to concentration c_0 the average speed is hardly influenced by the proximity of other vehicles, and that from there on, as concentration increases, the average speed declines approximately linearly. Equation (A.36) does not apply to conditions within queues.

---

[18] The distinction between "single" and "multivehicle" is based on what can be seen and recorded, not on the notion that single-vehicle crashes are those that occur without interaction with other vehicles. Thus, for example, if a driver falls asleep and crashes into another vehicle it is a multivehicle crash but if he wakes up at the last moment and manages to run off the road, it is a single-vehicle crash.

[19] This simple assumption is plausible for same-lane crashes and perhaps for adjacent-lane crashes (because $1/\overline{h} = \text{Flow}$).

[20] Averaging is over distance, not time.

[21] See, e.g., Transportation Research Board (2000) and Brilon (2011).

Because $\bar{v}$ is a function of "c," the "f" in (A.35) is a function of "c" only. To see what this $f(c)$ looks like a numerical example may be of use. The computations are in Fig. A.4.

For the "c" in column B the $\bar{v}$ is computed in column C by (A.36) using the parameters in C2:C3. The Highway Capacity Manual (2000, Exhibit 13-2, upper curve) implies that for a freeway on which $v_0 = 120$ km/h, $c_0 \approx 11$ vehicles/lane-km, $c_b \approx 28$ vehicles/lane-km, and $\beta \approx 2$ [(km/h)/(vehicles/lane-km)]. Using "c" and "$\bar{v}$" and the chosen parameters for "p" and "r" in E2:E3 these factors of (A.35) are computed in columns D and E; their product is the "f" in column F. The shape of $f(c)$ for the chosen functions and parameter values is in Fig. A.5.

	B	C	D	E	F	G
2		β= 2		α_m= 2		
3		V_0= 120		γ_m= 3		
4						
5	c [veh/lane-km]	$\bar{v}$ [km/h]	$c^{1+\alpha_m}$	$\bar{v}^{\alpha_m+\gamma_m}$	f	q=c$\bar{v}$ [vph]
6	0.0	120	0.0	0.0249	0.0000	0
7	0.5	120	0.1	0.0249	0.0000	60
8	1.0	120	1.0	0.0249	0.0002	120
60	27.0	88	19683.0	0.0053	0.8656	2376
61	27.5	87	20796.9	0.0050	0.8638	2393
62	28.0	86	21952.0	0.0047	0.8606	2408

Fig. A.4 Computations for multivehicle accidents

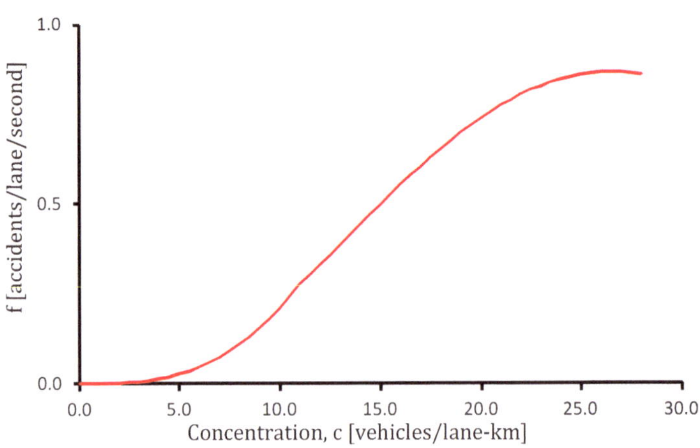

Fig. A.5 Multivehicle accident frequency as a function of concentration when $\alpha_m = 2$ and $\gamma_m = 3$

We usually have data about traffic flow (q), not about concentration (c). This is why for SPF development the functional dependence of "f" on "q" is of interest.[22] Because $q = \frac{1}{h} = c\overline{v}$,

$$f = \frac{q}{\overline{v}} \times q^{\alpha_m} \times (\overline{v})^{\gamma_m} = q^{1+\alpha_m} (\overline{v})^{\gamma_m - 1} \tag{A.37}$$

Here $\overline{v}$ as the implicit function of q:

$$\overline{v}(q) = \begin{cases} v_0 & \text{when } 0 < q < q_0 \\ v_0 - \beta\left(\frac{q}{\overline{v}(q)} - \frac{q_0}{v_0}\right) & \text{when } q_0 \le q < q_b \end{cases} \tag{A.38}$$

Up to $q_0 \equiv c_0 v_0$ traffic is assumed to move at the free-flow speed. When $c_0 = 11$ vehicles/km and $v_0 = 120$ km/h then $q_0 = 1{,}320$ vehicles/h and at $c_b \approx 28$ vehicles/lane-km, $q_b \approx 2{,}400$ vehicles/lane-hour, and $v_b \approx 86$ km/h. Column G in Fig. A.4 contains the values of "q" that correspond to "c" and "$\overline{v}$" in columns B and C. The shape of $f(q)$ for the chosen functions and parameter values is in Fig. A.6.

Single-Vehicle Accidents

The denser the traffic the less likely it is for a stray vehicle to run off the road without first hitting another car. To formulate a plausible function for "p" let p_0 be the probability of a vehicle to run out of control. That is, $p = p_0$ when $c \to 0$ and the

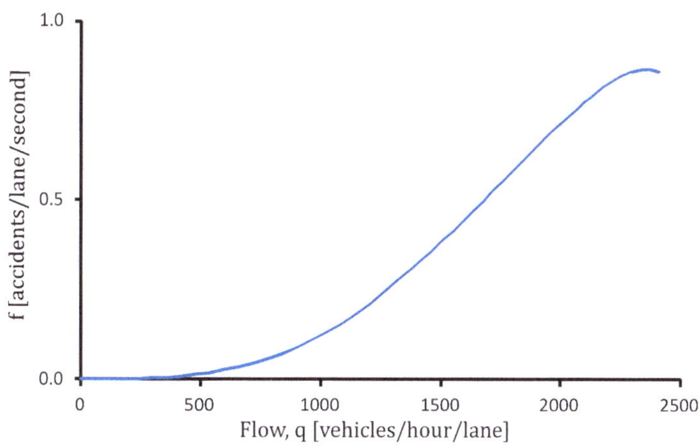

Fig. A.6 Multivehicle accident frequency as a function of traffic flow when $\alpha_m = 2$ and $\gamma_m = 3$

[22] In curve-fitting the traffic flow data is often that of AADT. However, if there is to be a meaningful causal relationship between crash frequency and traffic flow, q must refer to a period measured in minutes, not more than a few hours.

average headway is very large. As "c" increases, the chance of an errant vehicle to encounter another vehicle increases. As before, the probability of such an encounter is taken to be proportional to $1/\left(\overline{h}\right)^{\alpha_s}$, $\alpha_s \geq 0$. Let $\overline{h}_{\mathrm{b}}$ denote the average headway when $c = c_{\mathrm{b}}$ and assume that at that headway there are very few single-vehicle accidents. If so, a plausible function for p might be

$$p = p_0\left[1 - \left(\frac{\overline{h}_{\mathrm{b}}}{\overline{h}}\right)^{\alpha_s}\right] = p_0\left[1 - \left(\frac{c\overline{v}(c)}{c_{\mathrm{b}}v_{\mathrm{b}}}\right)^{\alpha_s}\right] \tag{A.39}$$

With this,

$$f(c) \propto c \times p \times r = cp_0\left[1 - \left(\frac{c\overline{v}(c)}{c_{\mathrm{b}}v_{\mathrm{b}}}\right)^{\alpha_s}\right](\overline{v})^{\gamma_s} \tag{A.40}$$

The computations of the ordinates "f" for arguments of the abscissa "c" are in Fig. A.7 and the corresponding $f(c)$ and $f(q)$ for the assumed functions and parameter values are in Figs. A.8 and A.9.

Figures A.6 and A.9 are the visual embodiment of those factors of the model equation which might represent the influence of traffic flow (q) on multivehicle and on single-vehicle accidents. These are not algebraic expression selected from a repertoire of simple and fully formed functions commonly used in curve-fitting. Rather, they are fairly complex expressions that are built up gradually and which represent explicitly stated postulates, assumptions, and data-based empirical equations.

The postulate is the conceptualization in (A.34). The assumptions are about how the probability to be in an accident and the probability to report that accident depends on various causal factors. The empirical element was about how the average speed depends on concentration.

	B	C	D	E	F	G
2			p_0= 1			
3		β= 2	α_s= 0.5			
4		V_0= 120	γs= 3			
5						
6	c [veh/lane-km]	v [km/h]	p	r	f	q [vph]
7	0	120	1.000	0.31	0.00	0
8	1	120	0.777	0.31	0.24	120
9	2	120	0.684	0.31	0.43	240
33	26	90	0.014	0.13	0.05	2340
34	27	88	0.007	0.12	0.02	2376
35	28	86	0.000	0.12	0.00	2408

Fig. A.7 Computations for single-vehicle accidents

Fig. A.8 Single-vehicle accident frequency as a function of concentration when $\alpha_s = 0.5$ and $\gamma_s = 3$

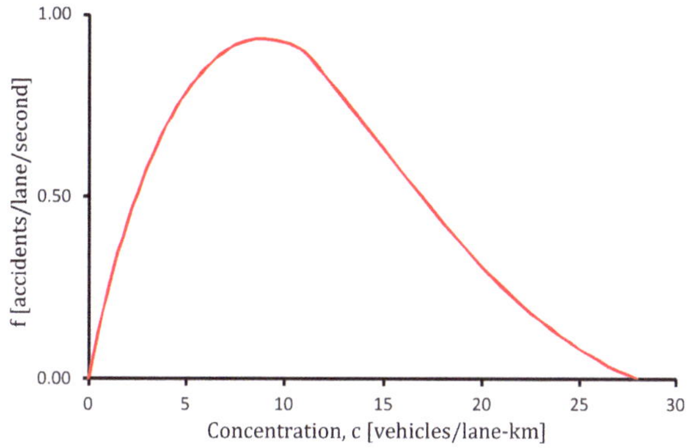

Fig. A.9 Single-vehicle accident frequency as a function of traffic flow when $\alpha_s = 0.5$ and $\gamma_s = 3$

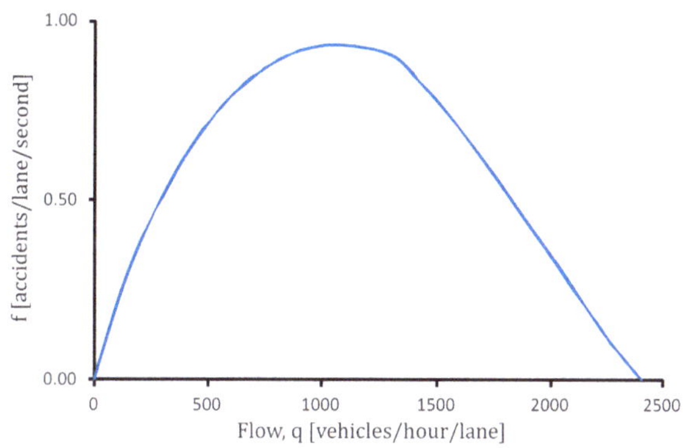

The main assumptions were two. One was about how the probability of a crash depends on the average headway. The other was about how crash severity and thereby the proportion of crashes that get reported depends on the average speed. These are plausible, simplistic, and provisional building blocks. Different plausible building blocks could have been used. Which building block (assumption) better fits the available data is an empirical question. The hope is that within the conceptualization $f = c \times p \times r$, research about crash generation and testing against data will lead to an improved representation of how "p" and "r" depend on various factors.

The empirical equations based on data-based research are about the relationship between average speed and concentration (A.36) or, equivalently, traffic flow (A.38). These make "f" into fairly complicated functions of "c" or of "q." Such functional forms cannot be discovered by curve-fitting. While one may stumble upon a workable approximation, the diffuseness of the available data clouds causes "goodness of fit" to be an ambiguous and fallible guide to functional form. A conceptual framework such as that proposed here may be a better guide for choosing a good functional form when fitting a curve to data.

The probability of a crash to occur for single and multivehicle crashes was assumed to be dissimilar. This is in line with what extant data suggest.[23] It follows that, if possible, separate models should be used to fit these two crash types. If, as is the habit, "total" (single + multivehicle) crashes were to be fitted the result would be a superposition of the curves in Fig. A.6 and A.9. It would take a complex function indeed to replicate all the resulting inflections and transitions. Simple functions are unlikely to well replicate the more complex reality.

For the empirical and assumed functional dependencies to make much sense, the flow data must be for time periods measured in minutes or at most a few hours. One cannot expect the conceptual framework to be of guidance when traffic flow data pertain to an average for a typical day of the year (AADT). This issue has been called "argument-averaging" by Mensah and Hauer (1998). Flow averaging is not the only source of averaging bias. The conceptual framework developed here is also vulnerable to biases due to the use of average speed and average headway.

By marrying some elementary traffic flow theory with the empirical knowledge about how drivers choose their speed of travel, and by making a few assumptions, it was possible to erect a conceptual framework for examining how crash frequency might depend on traffic concentration or on traffic flow. As the figures show, this conceptual framework can produce functional relationships that are not contradicted by available data; it may provide an explanation for what is observed and guidance for what function should be chosen for parameter estimation. From data about accidents and traffic flow for short time periods the α's, β's, and γ's could be estimated.

The proposed conceptual framework does not amount to a theory; a theory is an explanation of reality that has been sufficiently tested so that most scientists agree on it. The hope is that it opens a fruitful direction of investigation that could, eventually, help a theory to emerge and causal interpretations to become trustworthy.

[23] See, e.g., Ivan et al. (1999) and Jonsson et al. (2009).

Appendix K: The "Bump Function"

Logic suggests and experience shows that simple algebraic expressions often do not fit data in the entire domain of a variable; that there are regions where the simple function needs to be bumped up or down. A "bump function" (BF) can be used to multiply the fitted values in that region. To make sure that after such multiplication the model equation has no discontinuities, at the edges of the range the bump function should be 1.

Let X_l denote the lower limit of the range of variable X to which the Bump Function is to apply and X_u its upper limit. Define

$$t \equiv \frac{X - X_l}{X_u - X_l} \quad \text{when } X_l \leq X \leq X_u \tag{A.41}$$

A convenient Bump Function (BF) is

$$\begin{aligned} \mathrm{BF} = & \left[1 + t^{\beta_1}(1-t)^{\beta_2}\right] \text{if the fitted values are too small or} \\ & \left[1 - t^{\beta_1}(1-t)^{\beta_2}\right] \text{ if the fitted values are too large} \\ & \text{with } \beta_1 > 0 \text{ and } \beta_2 > 0 \end{aligned} \tag{A.42}$$

The shape of this function for selected parameter values is shown in Fig. A.10.

To get an average multiplier of, say, 0.9 one has to use the function with the minus sign in (A.42) and select the β's so that the area between the $\mathrm{BF} = 1$ line and the curve is about 0.1.[24]

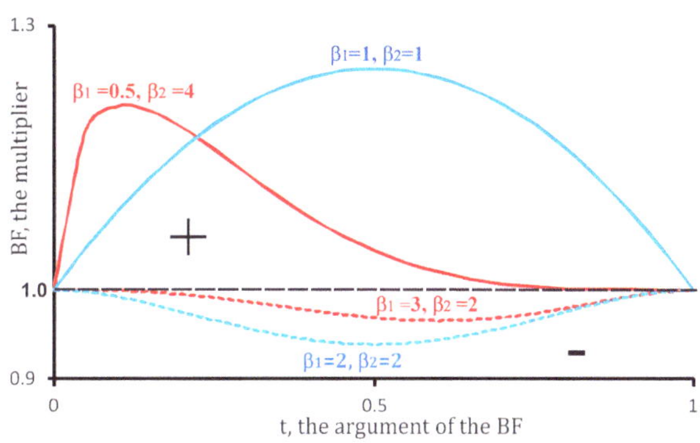

Fig. A.10 The Bump Function

[24] The area between the $\mathrm{BF} = 1$ line and a curve is the, so-called, Beta function. For β_1 and β_2 integers the area $\frac{(\beta_1-1)!(\beta_2-1)!}{(\beta_1+\beta_2-1)!}$. For non-integer values the area is $\frac{\Gamma(\beta_1)\Gamma(\beta_2)}{\Gamma(\beta_1+\beta_2)}$ and can be computed in Excel by using combinations of Exp(GAMMALN(.)).

Appendix L: Elasticity and CMFs for Multiplicative Single-Variable Functions

Claim 1: For multiplicative single-variable model equations elasticity is a function of one variable only and therefore model equations of this family cannot represent interaction.

Proof: For the general model equation $E\{\mu\} = f(X_1, X_2, \ldots, \beta_0, \beta_1, \ldots)$ the elasticity of "f" with respect to variable X_i is defined as $\varepsilon_{f,X_i} \equiv \frac{\partial f}{\partial X_i}\frac{X_i}{f}$. The partial derivative of $\beta_0 f_1(X_1, \boldsymbol{\beta}_1) f_2(X_2, \boldsymbol{\beta}_2)\ldots$ with respect to X_i is $\frac{\partial f}{\partial X_i} = \frac{f}{f_i(X_i,\boldsymbol{\beta}_i)}\frac{\partial f_i(X_i,\boldsymbol{\beta}_i)}{\partial X_i}$. The elasticity of "$f$" with respect to X_i is $\varepsilon_{f,X_i} \equiv \frac{\partial f}{\partial X_i}\frac{X_i}{f} = \frac{f}{f_i(X_i,\boldsymbol{\beta}_i)}\frac{\partial f_i(X_i,\boldsymbol{\beta}_i)}{\partial X_i}\frac{X_i}{f} = \frac{\frac{\partial f_i(X_i,\boldsymbol{\beta}_i)}{\partial X_i} X_i}{f_i(X_i,\boldsymbol{\beta}_i)}$ which is a function of only X_i.

Claim 2: For multiplicative single-variable model equations CMFs are constants and model equations of this family cannot represent interaction.

Proof: $\mathrm{CMF}(X_{i,a}, X_{i,b}) \equiv \frac{\beta_0 f_1(X_1,\boldsymbol{\beta}_1) f_2(X_2,\boldsymbol{\beta}_2), \ldots f_i(X_{i,a},\boldsymbol{\beta}_i), \ldots}{\beta_0 f_1(X_1,\boldsymbol{\beta}_1) f_2(X_2,\boldsymbol{\beta}_2), \ldots f_i(X_{i,b},\boldsymbol{\beta}_i), \ldots} = \frac{f_i(X_{i,a},\boldsymbol{\beta}_i)}{f_i(X_{i,b},\boldsymbol{\beta}_i)}$

which, given values for $X_{i,a}$ and $X_{i,b}$, is a constant.

Conclusion: For multiplicative single-variable model equations elasticity depends only on X_i and for given values of $X_{i,a}$ and $X_{i,b}$ the CMF is a constant.

Appendix M: Interaction Terms for Additive Linear and Multiplicative Power Models

Following Jaccard and Turrisi (2003) consider a linear additive model such as $Y = \beta_0 + \beta_1 X_1 + \beta_2 X_2$ in which X_2 is a moderator variable that affects β_1 via $\beta_1 = \alpha_0 + \alpha_1 X_2$. Substituting, $Y = \beta_0 + (\alpha_0 + \alpha_1 X_2)X_1 + \beta_2 X_2 = \beta_0 + \alpha_0 X_1 + \beta_2 X_2 + \alpha_1 X_1 X_2$. The effect of the assumed linear interaction is to add a product term and to change the regression coefficient of X_1.

Suppose now that X_1 is also a moderator variable and affects β_2 through $\beta_2 = \gamma_0 + \gamma_1 X_1$. Substituting, $Y = \beta_0 + (\alpha_0 + \alpha_1 X_2)X_1 + (\gamma_0 + \gamma_1 X_1)X_2 = \beta_0 + \alpha_0 X_1 + \gamma_0 X_2 + (\alpha_1 + \gamma_1)X_1 X_2$. The effect of the assumed double linear interaction is to add a product term and to change the regression coefficients of both X_1 and X_2.

Consider now the multiplicative model $Y = \beta_0 \times {X_1}^{\beta_1} \times {X_2}^{\beta_2}$ in which, again X_2 is a moderator variable that affects β_1 via $\beta_1 = \alpha_0 + \alpha_1 X_2$. Substituting, $Y = \beta_0 \times {X_1}^{\alpha_0 + \alpha_1 X_2} \times {X_2}^{\beta_2} = \beta_0 \times {X_1}^{\alpha_0} \times {X_2}^{\beta_2} \times {X_1}^{\alpha_1 X_2}$. The effect of the assumed linear interaction is to change β_1 into α_0 and to add the term ${X_1}^{\alpha_1 X_2}$.

Suppose now that X_1 is also a moderator variable that affects β_2 through $\beta_2 = \gamma_0 + \gamma_1 X_1$. Substituting, $Y = \beta_0 \times {X_1}^{\alpha_0 + \alpha_1 X_2} \times {X_2}^{\gamma_0 + \gamma_1 X_1} = \beta_0 \times {X_1}^{\alpha_0} \times {X_2}^{\gamma_0} \times {X_1}^{\alpha_1 X_2} \times {X_2}^{\gamma_1 X_1}$. The effect of the assumed double linear interaction is to add the factors ${X_1}^{\alpha_1 X_2}$ and ${X_2}^{\gamma_1 X_1}$ to change the regression coefficients of both X_1 and X_2.

Consider now the multiplicative model $= \beta_0 \times {X_1}^{\beta_1} \times {X_2}^{\beta_2} \times \beta_{\text{Terrain}}$ in which $\beta_{\text{Terrain}} = 1$ when Terrain = Flat, β_{Rolling} when Terrain = Rolling, and $\beta_{\text{Montainous}}$ otherwise. In this case the influence of the moderator variable X_2 could be captured simply by, say, $\beta_{\text{Terrain}}(1 + \alpha_{\text{Terrain}} X_2)$ as in Sect. 10.8.

References

Brilon W, Lohoff J (2011) Freeway flow models. In: 90th Annual meeting of the Transportation Research Board, Washington, DC

Freund FE, Walpole RE (1987) Mathematical statistics, 4th edn. Prentice-Hall, New York

Gerlough DL (1955) Use of Poisson distribution in highway traffic. The ENO Foundation, Saugatuck. http://ntl.bts.gov/lib/26000/26800/26814/use_of_poisson_distribution_in_highway_traffic.pdf

Guo G (1996) Negative multinomial regression models for clustered event counts. Sociol Methodol 26:113–132

Hauer E (2001) Overdispersion in modeling accidents on road sections and in Empirical Bayes estimation. Accid Anal Prev 33:799–808

Ivan J, Pasupathy R, Ossenbruggen P (1999) Differences in causality factors for single and multi-vehicle crashes on two-lane highways. Accid Anal Prev 31:695–704

Jaccard J, Turrisi R (2003) Interaction effects in multiple regression, 2nd ed. Sage University papers on Quantitative Applications in the Social Sciences, 07-072. Sage, Thousand Oaks

Jonsson T, Lyon C, Ivan J, Washington S, van Schalkwyk I, Lord D (2009) Differences in the performance o safety performance functions estimated for total crash count and for crash count by crash type. Transp Res Rec 2102:115–123

Mensah A, Hauer E (1998) Two issues of averaging in multivariate modelling. Transp Res Rec 1635:37–43

Nicholson AJ (1985) The variability of accident counts. Accid Anal Prev 17:47–56

Nicholson AJ (1986) The randomness of accident counts. Accid Anal Prev 18:193–198

Nicholson AJ, Wong YD (1993) Are accidents Poisson distributed? A statistical test. Accid Anal Prev 25(1):91–97

Quine MP, Seneta E (1987) Bortkiewicz's data and the law of small numbers. Int Stat Rev 55 (2):173–181

Transportation Research Board (2000) Highway capacity manual. National Research Council, Washington, DC

von Bortkiewitcz L (1898) Das Gesetz der Kleinen Zahlen. Tebner, Leipzig

Index

E. Hauer, *The Art of Regression Modeling in Road Safety*,
DOI 10.1007/978-3-319-12529-9

MIX
Papier aus verantwortungsvollen Quellen
Paper from responsible sources
FSC® C105338

If you have any concerns about our products,
you can contact us on
ProductSafety@springernature.com

In case Publisher is established outside the EU,
the EU authorized representative is:
Springer Nature Customer Service Center GmbH
Europaplatz 3, 69115 Heidelberg, Germany

Printed by Libri Plureos GmbH
in Hamburg, Germany